ROUTLEDGE LIBRARY EDITIONS: ENERGY ECONOMICS

Volume 20

URANIUM

URANIUM
A Strategic Source of Energy

MARIAN RADETZKI

Routledge
Taylor & Francis Group

LONDON AND NEW YORK

First published in 1981 by Croom Helm Ltd

This edition first published in 2018
by Routledge
2 Park Square, Milton Park, Abingdon, Oxon OX14 4RN

and by Routledge
711 Third Avenue, New York, NY 10017

Routledge is an imprint of the Taylor & Francis Group, an informa business

British Library Cataloguing in Publication Data
A catalogue record for this book is available from the British Library

ISBN: 978-1-138-10476-1 (Set)
ISBN: 978-1-315-14526-6 (Set) (ebk)
ISBN: 978-1-138-50095-2 (Volume 20) (hbk)
ISBN: 978-1-138-50111-9 (Volume 20) (pbk)
ISBN: 978-1-315-14412-2 (Volume 20) (ebk)

Publisher's Note
The publisher has gone to great lengths to ensure the quality of this reprint but
points out that some imperfections in the original copies may be apparent.

Disclaimer
The publisher has made every effort to trace copyright holders and would welcome
correspondence from those they have been unable to trace.

Uranium

A STRATEGIC SOURCE OF ENERGY

MARIAN RADETZKI

CROOM HELM LONDON

British Library Cataloguing in Publication Data

Radetzki, Marian
    Uranium. - (Croom Helm commodity series).
    1. Uranium industry
    2. Uranium - Political aspects
    I. Title
    338.4'7'6692931091812   HD9539.U7

    ISBN 0-7099-0340-5

Typeset by Jayell Typesetting · London
Printed and bound in Great Britain
by Photobooks (Bristol) Ltd · Bristol

CONTENTS

TABLES

CHARTS

EDITORIAL STATEMENT

This book, *Uranium* by Marian Radetzki, is the first in a new series entitled the Croom Helm Commodity Series. The series has as its objective the advancement of the understanding of issues relating to the production and marketing of the major primary commodities. Many volumes in the series will concentrate on analysing the essential properties, production background, current production processes, marketing and prospects of a single commodity. Several will also deal with broader questions such as commodity pricing, international commodity control and commodity trade. A further volume will bring a fresh approach to the subject of the plantation — its background, different forms and its future role as a social and economic unit. Although differing markedly in content, the volumes will share a similar form and direction and it is hoped that they will be useful for reference purposes. Each of the volumes will also set the subject of commodity production in the context of the changing international economic environment and it is hoped that as a result some light will be shed on the future profile of commodity production and the problems which will continue to face both the producers and the consumers of commodities.

Contributors to the series come from a variety of background. They share, however, established reputations within their chosen fields, and we are confident that this will be clearly reflected in the quality of each volume in the series. The series should appeal to anyone interested in commodity issues. It will be of interest as well to students of the North-South dialogue, since by adding to our understanding of the problems associated with the production and marketing of commodities we may draw nearer to their solutions.

Fiona Gordon-Ashworth
Commodities Editor

FOREWORD

During the past three years, a major focus of the research efforts at the Institute for International Economic Studies, University of Stockholm, has been on issues related to the functioning of the international economic system. These activities have been supported by generous grants from the Bank of Sweden Tercentenary Fund. Marian Radetzki's study constitutes part of this research programme.

Dr Radetzki focuses on the instability of the international uranium market during the 1970s, and in particular on the uranium price explosion of 1974/5. A striking result of the study is that instability in the uranium market is related more to the shifts in government policies and regulations — in uranium-producing, uranium-consuming and oil-producing countries — than to actions undertaken by profit-orientated firms. This result in no way contradicts the fact that 'ordinary' market forces like demand and supply or cartel arrangements among firms determine the magnitude of the price responses to government measures.

The study points to the importance of carefully analysing the behavioural patterns of governments and government-regulated firms, and of incorporating the results of such analyses into economic research, for instance in the field of international commodity markets.

The major role of the Institute is to provide a suitable intellectual atmosphere and physical facilities for independent scholars. They themselves are solely responsible for the results of their research. This, of course, is also true in the case of the present study.

Assar Lindbeck
Director of the Institute for International Economic Studies
University of Stockholm

ACKNOWLEDGEMENTS

For almost ten years I have devoted most of my professional attention to various aspects of mineral economics. My interest in uranium was initially aroused through an assignment by the Westinghouse Corporation to study certain aspects of the market for this mineral. I found the economics of uranium so fascinating that I decided to expand the scope of my investigation and to pursue it on my own after the assignment had been completed. The result is the present book.

Most of the work was conducted in 1979 and early 1980. I have benefited from visits to and discussions with the Uranium Institute in London, the International Atomic Energy Agency in Vienna and Euratom's Supply Agency in Brussels. Valuable comments on various draft versions of the text have been received from Ake Blomqvist, Alf Carling, Fiona Gordon-Ashworth, Ake Sundstrom, Lars Tallbacka, Carl Van Duyne and a few friends in the international mining industry who prefer to remain anonymous. Bengt Gembel at the Studsvik Library in Sweden has made a great contribution by sorting out and supplying me with vast amounts of documentation. The Institute for International Economic Studies at the University of Stockholm has offered a stimulating atmosphere and excellent facilities for the work. Anita Oxenham has been tireless in typing endless manuscripts and smoothing out problems of style. The Bank of Sweden Tercentenary Fund, finally, provided generous financial support without which this work could not have been undertaken.

Marian Radetzki

1 THE ISSUES

This book deals with a highly unpredictable and volatile international commodity market. Though the subject coverage is limited to uranium, large parts of the analysis have a more general applicability to international commodity market problems. As will appear from the following, much of the instability experienced in uranium has had its origin in political rather than economic action, although inelasticities in demand and supply, the organisation of the uranium industry and the dominant influence of close substitutes have also contributed to the erratic performance of uranium prices.

The focus is on commercial, as distinct from military, uranium uses. A non-military market for uranium developed in the late 1960s in response to the demand for fuel for the growing number of nuclear reactors coming into use in various countries. From 1968 until the end of 1973, uranium prices remained depressed and relatively stable. Spot-delivery quotations varied within a range of \$5.50-7 per 1b U_3O_8,[1] with long-term contract prices executed at roughly corresponding levels.

In 1974, a drastic change began to take place. In an uninterrupted rise, spot prices increased more than sixfold, or from \$7 late in 1973 to above \$40 in mid-1976; they have remained at this high level until the present. The price development was not limited to spot transactions. Futures prices in contracts signed at different times moved more or less in parallel with the spot price changes.

The price figures given above are in nominal dollars. But even if adjustment is made for inflation, the change was extraordinary. Between 1973 and 1976 the uranium price in constant dollars[2] increased almost fivefold, and remained at that high level for about two years. In 1978 and 1979, however, there have been substantial falls in the real, as distinct from the nominal, uranium price quotation.

The price history of uranium is conveniently depicted in Chart 1.1. In anticipation of the detailed analysis in Chapter 2, it may be noted that price changes as large and sudden as the one which affected uranium are indeed very rare in the modern history of international commodity markets.

The extreme price performance of uranium provides much of the rationale for writing the present book. A major portion of the following text is devoted to analysing several interrelated factors, which emerged

Chart 1.1: Uranium Prices, Spot Deliveries, as Reported by NUEXCO Quarterly Averages

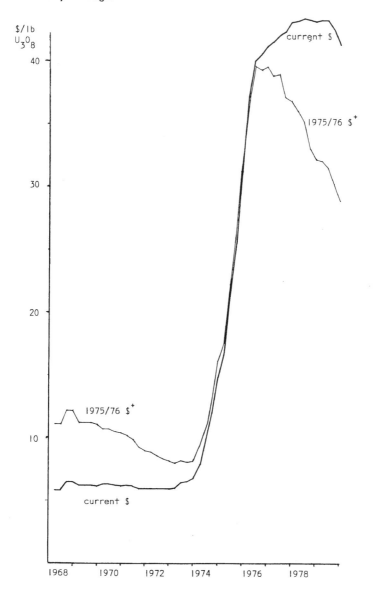

Note: [a]Current dollars deflated by the general dollar GDP deflator for the entire OECD area. The quarterly GDP deflator data have been obtained through linear approximations of annual figures (see Table 2.3).

within a short time, and which we see as primarily responsible for the uranium price changes.

In the process of explaining the price movements, the book provides a comprehensive analysis of the international uranium market, from its inception in the late 1960s up to the present. The issues are treated in the following order. The next two chapters set the scene. In Chapter 2 the price systems and price developments in the uranium market are explored in detail. The price movements for major commodities in international trade since 1920 are also studied, in order to substantiate the claim that the 1973-6 uranium price explosion has few, if any, precedents. In Chapter 3 we move back in time and document the circumstances under which uranium mining emerged, blossomed and declined as a result of the fast-growing and then equally fast-declining military demand. This account is a necessary prerequisite for a proper understanding of the depressed state of the uranium industry between 1968 and 1973, which resulted from the speedy curtailment and subsequent discontinuation of military procurement.

Chapter 4 contains a broad-based conventional analysis of the conditions which are normally instrumental in shaping price developments in minerals markets. The analysis includes a survey of consecutive nuclear power growth forecasts and of the ensuing uranium requirements,[3] a scrutiny of the adequacy of mineral reserves and production capacities for the satisfaction of current and prospective uranium needs, and a comparison of uranium output at different times with the simultaneous requirements of reactors in operation. It then explores production costs and establishes the price level necessary to attract investments in new production capacity to satisfy future needs. The results of these analyses, supplemented by a survey of price expectations formulated in the early 1970s, suggest that there was nothing in the ordinary market circumstances from which the impending price explosion could be anticipated.

Several of the following chapters detail the set of factors which were jointly responsible for the drastic change in the uranium market. Chapter 5 illuminates how political decisions regarding uranium reprocessing and enrichment contracting contributed to the price rise by boosting demand and locking it at a level above requirements. Chapter 6 considers the impact of the rising oil price on the uranium market. The analysis suggests that the major causal relationship was through the general increase in electricity prices, leading to the emergence of temporary excess profits (quasi rents) in nuclear electricity generation which uranium producers could appropriate. In Chapter 7, after scrutinising the structure of the uranium industry and concluding that it is amenable to monopolistic collusion, we analyse in detail the actual operations of the international uranium cartel from 1973 onwards. Chapter 8 discusses several further factors which also appear to have contributed to the upward move in uranium prices.

The concluding Chapter 9 summarises and synthesises the findings reached in earlier pages. It also discusses the extent to which the uranium experience can be applied to other international commodity markets. Finally, it provides an outlook for the future by considering the circumstances likely to influence the uranium market in coming years.

These, in brief, are the contents of the book. Several issues concerning the nuclear industry in general are not taken up at all or are treated only tangentially. For instance, our analyses of the uranium market are entirely based on the fuel requirements of the light water reactor, and disregard other existing or planned reactor types. Table 1.1 details the expected subdivision of the Western world[4] nuclear capacity between alternative reactors in 1980 and 2000, and indicates their respective uranium consumption patterns. Given the heavy dominance of the light water reactor through the present century, the limited differences in uranium consumption between most non-breeder reactor types and the unimportance of the breeder through this century, the above simplification does not distort our conclusions in any noticeable way.

Table 1.1: Forecast Distribution among Different Reactor Types and their Respective Uranium Consumption Patterns

	Forecast Capacity in Percentage of Western World Capacity		Annual Requirements of Natural Uranium lb U_3O_8 per MWe
	1980	2000	
Light water	87	90	440
Graphite gas	4.1	0.2	780
Advanced gas	4.1	0.6	430
Heavy water	4.8	7.6	440
High temp., gas cooled	0	0.6	350
Breeder	0	1.0	5

Source: OECD, *Uranium Resources, Production and Demand* (OECD, Paris, Dec. 1975 and Dec. 1977).

While elaborating in considerable detail on the sharp reductions in nuclear growth forecasts which have occurred through the 1970s, and on the ensuing consequences for future uranium demand, we do not attempt to provide any account of the factors underlying this change — i.e. the safety problems in nuclear energy and the popular disaffection with nuclear power of recent years — since these issues are discussed adequately elsewhere.[5]

The cost of investment in nuclear reactors, though very important for the nuclear industry, is also not central to our theme, and is therefore discussed only in passing, when the economics of nuclear electricity are compared with those of coal- and petroleum-based electricity generation.

Our analysis focuses on the short, medium and long term, the latter bounded by a time horizon of fifteen to twenty years. We do not aim to comment in detail on the very long term in discussing the sufficiency or otherwise of uranium mineral resources.

There are several general points which it is useful to cover at this stage, even though they will be treated in greater detail later in the text. The information contained in Chart 1.2 may be helpful in this context.

The first clarification aims at providing useful reference figures regarding the global supply of and requirements for uranium. In 1978, Western world uranium output amounted to 45,000 short tons U_3O_8. At a price of \$40/lb this represented a total value of \$3,600 million. In 1973, production was only 26,000 short tons, and with a price of \$7/lb, the value was no higher than \$360 million, i.e. only one-tenth of the 1978 figure. The 1978 requirements of uranium to fuel reactors were estimated at 34,000 short tons, i.e. 12,000 short tons below the production figure in that year. Excessive production has prevailed through the 1970s, and this has resulted in fast-rising global inventories.

Chart 1.2 also provides a rough impression of the international trade flows in uranium. Three important consumer countries, i.e. the US, Canada and France, are self-sufficient on the basis of domestic production. Of the three, only Canada produces a substantial surplus which is available for export. All other major consumers are entirely dependent on imports for their requirements. South Africa, Niger and Gabon, with limited current domestic uranium needs, are, apart from Canada, the largest suppliers to the international market. Australia is expected to become a large-scale producer and exporter by the mid-1980s. There are no substantial differences between the prices of domestically and internationally traded uranium.

An additional important piece of information contained in Chart 1.2 is the heavy national concentration of supply. Four countries account for about 90 per cent of the entire Western world output. This is considerably more than in most other minerals. The four-country share is below 80 per cent for tin, 70 per cent for bauxite, and around 60 per cent for copper and petroleum. The high country concentration makes the uranium market particularly sensitive to disruptions through political measures undertaken by major producing countries.

Chart 1.2: Western World Production and Reactor Requirements of Uranium in 1978 (thousand short tons U_3O_8)

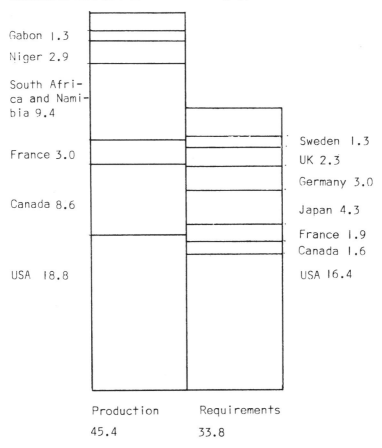

Gabon 1.3		Sweden 1.3
Niger 2.9		UK 2.3
South Africa and Namibia 9.4		Germany 3.0
France 3.0		Japan 4.3
Canada 8.6		France 1.9
		Canada 1.6
USA 18.8		USA 16.4

Production   Requirements
45.4   33.8

Note: Requirements based on installed reactor capacity and enrichment tails of 0.2 per cent (see Chapter 4).

Source: *The Balance of Supply and Demand 1978-1990* (The Uranium Institute, London, 1979).

Uranium supply is heavily concentrated not only at the national level, but also among corporate decision units. In the case of South Africa (excluding Namibia) and France, the overall supply is handled by single nation-wide marketing organisations. In addition, the French marketing agency maintains strong control over output originating in Gabon and Niger. The six largest corporate decision units account for about two-thirds of total Western world output. The degree of corporate concentration, though not exceptional among minerals industries, is certainly high enough to enable profitable producer collaboration in the management of supply.

To broaden the perspective, the figures for uranium output, price and value in 1978 are compared in Table 1.1 with those for three other minerals.

Table 1.2: Western World Production and Value of Petroleum, Iron Ore, Uranium and Silver in 1978

	Output (thousand short tons	Price ($ per short ton	Value of Output ($ million)
Petroleum	2,640,000	86	227,000
Iron ore	475,000	17	8,075
Uranium	45	80,000	3,600
Silver	9	160,000	1,440

Another clarification concerns the means by which natural uranium is transformed into nuclear reactor fuel. The predominant pattern is that uranium oxide (U_3O_8) is sold by uranium mine and mill enterprises directly to the utilities which operate nuclear reactors. The utilities then have the uranium processed into nuclear fuel elements through conversion, enrichment and fuel fabrication on a toll basis (i.e. by outside firms, while the utilities themselves maintain legal possession of the material). The enrichment service is entirely in public ownership, and was until very recently a virtual monopoly of the US government. Conversion and fuel-fabrication services are provided by private firms. Although the conversion and fabrication industries are heavily concentrated, it appears that both are working under competitive conditions. The same was true for uranium-mining and -milling until 1973, when the international uranium cartel established itself firmly in the uranium concentrate market. Up to Chapter 7, which explicitly treats the cartel issue, a fair degree of competition among the uranium producers is implicitly assumed in the analysis.

A few definitions and terms need to be established. Throughout this book the words *world* and *global* refer to the Western world only, i.e. they exclude the centrally planned economies of Eastern Europe and Asia. This is because few data are available on these countries' uranium activities, and because, apart from enrichment, the centrally planned economies have very little uranium interchange with the rest of the world. All uranium prices are expressed in terms of US dollars per lb U_3O_8. All uranium quantities are given in short tons U_3O_8. For easy comparison with literature using other units, it may be noted that U_3O_8, uranium oxide, contains about 85 per cent U, uranium metal, and that one metric tonne of U corresponds to 1.3 short tons of U_3O_8.

In view of the US government's dominance in the development of nuclear technology, frequent reference is made to the authority in the US administration responsible for these matters. This authority has changed name and designation several times during the period covered by the present study. Until 1974, the relevant institution was the US Atomic Energy Commission (AEC). Between 1974 and 1977, the body dealing with the issues which are treated in the present book was called the US Energy Research and Development Administration (ERDA). Since that time the responsibilities have rested with the US Department of Energy (DOE).

Finally, a few words about style and approach. Though the book deals primarily with *economics*, we make it a point to avoid using the more complex formal tools of economic theory and of empirical economic analysis. There are two reasons for this approach. First, the desire is to present a text accessible to the educated layman without any formal training in economics. Second and more important, however, we are doubtful whether more formal theoretical or empirical analysis, for instance through the application of advanced oligopoly theory, or the use of econometric tools on our material, would have helped in obtaining additional or more reliable insight into the subject-matter.

When data are inadequate or uncertain, we find it instructive to provide numerical examples of the issues under scrutiny. These examples are then partly based on reasonable assumptions which fill in the data lacunae. The outcome of the exercises should be regarded as reasonable illustrations of real conditions and not as statements of fact.

In an attempt to get at least a feel for the relative market impact of the respective factors which are seen as the causes of the price explosion, the cumulative five-year change in supply or demand caused by each are quantified wherever possible. The five-year period supply and demand shifts which are quantified all include the latter half of the 1970s, but due to lack of data in some cases, the overlap between them is not complete.

Notes

1. Unless otherwise indicated, dollars refer to US dollars throughout the book. The final product sold by uranium mines is uranium concentrate, or yellowcake. Uranium prices are quoted per lb uranium oxide, U_3O_8, contained in the concentrate. The various processing stages through which uranium concentrate is transformed into nuclear fuel elements are discussed in detail in Chapter 4.

2. Except where otherwise specified, constant dollar figures are obtained by deflating nominal dollars by the general dollar GDP deflator for the entire OECD area. The reason for choosing this deflator is explained in Chapter 2.

3. A distinction is maintained throughout the analysis between uranium *requirements*, i.e. the quantities needed to fuel reactors or to satisfy delivery obligations in enrichment contracts, on the one hand, and *demand*, which also includes the (positive or negative) quantities for inventory change.

4. Western world is defined to include all countries except the centrally planned economies of Eastern Europe and Asia.

5. These issues are discussed, for instance, in G.M. Ferrara (ed.), *Atomic Energy and the Safety Controversy* (Checkmark Books, New York, 1978); P. Faulkner (ed.), *The Silent Bomb: a Guide to the Nuclear Energy Controversy* (Vintage Books, New York, 1977); and E.S. Rolph, *Nuclear Power and the Public Safety* (DC Heath, Lexington, 1979), to which the interested reader is referred.

2 URANIUM PRICES IN BOOM AND BUST

Introduction

This chapter is devoted to a study of commercial uranium prices from 1968, when a market for non-military uranium became established, to the present. The exceptionality of the uranium price performance provides a rationale for the space and detail afforded to our price analysis.

Price can be an important indicator of the conditions that prevail in a market, reflecting the strains and disruptions in demand and supply and the consequent impact on industries that depend for their livelihood on the commodity in question. One must be cautious, however, before attaching too great a significance to a series of price quotations, or to the development of price levels. Several issues are involved.

One important consideration in studying commodity price movements is the generality of a specific price series. The prices most widely quoted in the international sugar market, for instance, refer to marginal quantities only. Most international trade in sugar is transacted under long-term contracts at stable prices, which have little relationship to the highly volatile residual world market. Similarly, in zinc until 1976, a major proportion of overall output was sold at producer-set prices which seldom corresponded to the London Metal Exchange (LME) quotation. The price swings in the free sugar and zinc markets have had a limited significance, since they have applied to marginal quantities only. The major share of producers' output and consumers' needs has been satisfied at other, usually much more stable, price levels. The price instability in free markets like those for sugar and zinc is amplified by the marginality of such markets. Producers and consumers will be less reluctant to accept extremely high or low price levels for marginal quantities when a major share of their transactions takes place at more 'reasonable' prices.

Another important consideration in judging the significance of commodity price changes to the industries involved is the abruptness with which they occur. Substantial price movements can be absorbed without excessive strain when the change occurs gradually and is extended in time, by consumers substituting into other, cheaper materials, or by high-cost producers leaving the industry after having depreciated their capital stock. Sudden price shifts, on the other hand, to the extent that they cannot be absorbed by changing inventories, create far more severe problems in requiring fast and painful adjustments in the producing and consuming industries.

When judging the significance of a price change, one must also clarify the durability of the price level which emerges after the change. The shorter the period during which the new price remains, the less significance it will have for producers and consumers. Inventory variations will enable producers and consumers to avoid temporary troughs and peaks respectively by postponing their sales and purchases until more normal price levels re-emerge.

For markets where both current and futures prices are quoted, one can judge to some extent the expected durability of a price change by studying the extent to which futures prices are affected along with spot quotations. A short-run disturbance may lead to a sharp change in the spot market, but its impact on the futures market is bound to be weaker. A disturbance judged to have a lasting impact should influence current and futures prices in equal measure.

Commodity market prices are usually stated in nominal terms, i.e. in current dollars, pounds or francs. For periods of fluctuating exchange rates and fast inflation, it is necessary to transform the nominal price changes into real ones, in order to obtain a proper impression of their significance. This becomes particularly important when price developments over somewhat longer periods are considered. It may then be found that a strong price rise expressed in a depreciating currency may disappear when converted into real terms. Conversely, a stable nominal price level may imply a substantial deterioration in real terms under similar circumstances. In the following discussion of uranium prices, we devote special attention to the significance of available price quotations and of the changes in price levels that occurred in the period studied. We also widen the perspective by comparing the price changes affecting uranium in the 1970s with price developments for a great number of primary commodities in international trade at different times. This comparison permits us to judge whether the recent performance of uranium prices in the international market was exceptional or not.

The Magnitude and Generality of the Uranium Price Rise of the 1970s

There is no single world market exchange for uranium. Hence, uranium prices at a given point in time will vary depending on the geographical location and on the conditions under which the transaction takes place. In these circumstances, it is difficult to present general uranium price series over time. The problem is aggravated by the fact that the price levels of many transactions are not publicly reported. In what follows we try to disentangle the price issue as far as available evidence permits. Our problem for uranium is not unique: similar difficulties are encountered in most commodity markets,[1] even though in some (e.g. copper and lead) the LME quotation is widely accepted as a representative reference price, and used in contractual arrangements outside the Exchange.

Most uranium has been sold under long-term contracts, usually covering at least five years' deliveries, and spot transactions have constituted only a marginal share of total trade. Uranium contracts are private documents, and their contents are seldom published. Hence, the information on what such contracts contain is not comprehensive. In the US, contracts traditionally specified a base price and rules for escalation to protect the seller from the effects of inflation. Contracts with European customers up to the early 1970s frequently specified the delivery prices for the entire contract period.[2]

The price data are much more systematically available in the US than in the rest of the world. This is primarily due to the regular investigations of uranium contracts undertaken by the US Department of Energy (DOE) and its predecessors, and the 'exchange values' published monthly by the Nuclear Exchange Corporation (NUEXCO), a private brokerage firm which acts as agent for uranium transactions in the US.

The DOE data provide averages of prices for deliveries at different times, as stipulated in US contracts in existence at the time of enquiry. Thus, in the survey of January 1973, the average price was $7.10 for deliveries in the same year and $7.80 for deliveries in 1980. In the survey of January 1976, the average price was $10.70 for deliveries in the same year, and $14.35 for deliveries in 1980.[3] In the survey of January 1978, same-year delivery commitments averaged $17.40, while the price for deliveries in 1980 was estimated at $18.50.[4] The DOE data should provide a reasonable average figure for prices actually paid under contract deliveries at the time of each survey. On the other hand, they provide a misleading impression of average prices for deliveries in future years, since these averages are bound to be affected by contracts not yet written at the time of the survey. Also, since existing contracts at each point of time will have been entered into over a number of years in the past, the DOE average will be very slow in catching a sudden change affecting the prices of all new contracts.

The NUEXCO exchange values, in contrast, disregard the price arrangements of old contracts, and concentrate on the current state of the market. They express NUEXCO's judgement of price levels in current transactions involving either immediate or future deliveries.[5] Chart 1.1 presented the developments of spot prices between 1968 and 1979, as reported by NUEXCO. Table 2.1 complements some of the spot figures with prices for future deliveries contracted at the time of reporting. The table reflects the drastic change in the US uranium market in the mid-1970s, and indicates that the change affected spot and futures prices in equal measure. From July 1973 to July 1976, for instance, there was a sixfold increase in the spot price as well as in the contracted price for a four-year future delivery.

Table 2.1: NUEXCO Exchange Values for Spot and Future Deliveries of U_3O_8 ($/lb)

Report Date	Spot	July 1977	Delivery Date July 1978	July 1979	July 1980
July 1972	5.95	8.30	—	—	—
July 1973	6.50	8.65	9.30	9.90	10.50
July 1974	11.50	15.25	16.50	18.50	19.50
July 1975	24.70	28.35	30.20	33.35	35.55
July 1976	40.00	42.80	45.60	48.55	51.75
July 1977	42.25	42.25	44.95	48.10	51.45
July 1978	43.40	—	43.40	—	—
July 1979	42.70	—	—	42.70	—

Source: Various NUEXCO reports.

Uranium produced in the US has been exported, off and on, to Western Europe and Japan. On the other hand, in view of the embargo on the use of foreign uranium in US reactors in force between 1964 and 1977, and gradually relaxed thereafter,[6] the uranium market in the US has been somewhat insulated from the rest of the world. This insulation appears to have created, at times, a noticeable, though not overwhelming, difference in the price levels within and outside the US. In a 1974/5 US Bureau of Mines publication, it is noted that uranium prices have been lower outside the US, but that the trend has been towards international equalisation in price levels.[7] This tendency, partly due to US imports of foreign uranium in anticipation of a removal of the US embargo, can be seen in Table 2.2, which compares US and European future delivery prices, contracted between 1971 and 1973.

Table 2.2: US and European Uranium Prices for Future Delivery
($/lb U_3O_8)

Quotation Period	Delivery Period	US Price	European Price	European Price in Percentage of US Price
December				
1971	1974	6.70	5.00	75
June 1972	1976	7.70	6.90	90
1973	1977	8.25 – 9.50	6.90 – 9.60	84 – 101

Source: NUEXCO, various reports, and Euratom Versorgungsagentur, *Annual Reports*, 1972 and 1973.

For the period after 1974, available data suggest a relatively close correspondence between US and European quotations. Thus, European prices for deliveries in 1975 are reported to have risen from about $8 to some $17 in the course of 1974.[8] This is quite close to the developments in the US reported by NUEXCO. In the latter part of 1975, spot prices in Europe were reported to be about $25,[9] or somewhat lower than the US price level. In 1976, on the other hand, European spot prices were assessed at between $39 and $42 for deliveries from South Africa, Canada and Niger[10] (US spot prices in 1976 varied between $35 and $42), though another source indicates that some European spot transactions early in the year were as low as $30-35.[11] Price levels for long-term contracts outside the US are reported to have attained a base level of $30-35 in the first half of 1977, with price escalation to cover inflation for deliveries in later years.[12] This seems to be low compared to US futures prices agreed upon in the same period. On the other hand, prices of $42.70 are quoted in a contract between a Canadian producer and a Swedish utility, signed in June 1977, for deliveries up to 1981.[13] Spot prices from January 1978 to October 1979 both in Europe and in the US remained within a relatively narrow range of $42-44.[14] A fairly safe conclusion from this scanty non-US price evidence is that the price change in Europe from the early 1970s was of the same magnitude as that experienced in the US, and that prices reached about the same levels in Europe and the US in the late 1970s.

The significance of the price rise was further increased by widespread efforts of uranium producers to renegotiate old contracts that committed them to supply for long periods at prices far below the market levels recorded since 1976. Consumer acceptance of renegotiation appears to have been widespread.[15] Thus, it is reported that practically all uranium consumers in the European Community had agreed in the course of 1976 to such renegotiation with their overseas suppliers. As a result, the former prices of around $10 in contracts entered into between 1972 and 1974 were considerably increased, on average to some $16.[16] British Nuclear Fuels, for instance, renegotiated its contract for uranium supplies from Rossing Uranium from less than $10 for 1977 deliveries to about $13.[17] British and German utilities were also pressured in 1976 to accept a price increase from $7-8 to about $20-25 in a long-term supply contract with Nufcor of South Africa.[18] The Swedish State Power Board agreed to an increase in prices in a 1973 contract with Uranex for 1975-8 deliveries from some $7-10 per lb U_3O_8 to about $14-40.[19] Japanese buyers also yielded to producers' pressures to renegotiate. For instance, it is reported that eight Japanese utilities which had concluded long-term contracts for deliveries up to 1984 with a Canadian supplier, and paid $12 in 1975, had agreed to pay $21.45, or 79 per cent more, in 1976.[20] In an inter-governmental discussion, Japan agreed early in 1976 to renegotiate the price of 9,000 tons of uranium bought from Australia in 1972 at $8 per lb for future deliveries.[21] Along with British and German utilities, the Japanese also accepted a renegotiation of prices for deliveries fron Nufcor of South Africa. The new Japanese price was reported at between $25 and $30, i.e. even higher than that accepted by Nufcor's European clients.[22] To explain fully the substantial variety of outcomes reached in the renegotiations discussed here, one would have to look into such factors as the exact time when each was concluded, political or other ties between the negotiating parties, and the renegotiation strategies adopted by different producers.

Around 1975, uranium suppliers became less willing to enter into long-term contracts with specified prices and/or escalation clauses over the entire contract period. In some cases they pressed the buyers to provide finance for mines to be developed as a condition for long-term supply contracts. In others, the long-term contracts specified prices for early deliveries only, and provided for price negotiations for later years. In more recent years, so-called market price contracts have become common.[23] The price rules in these contracts vary considerably, but they commonly require the buyer to pay the 'world market price' at the time of delivery, or an escalating base price, whichever is higher. NUEXCO quotations are presumably used to determine the 'world market price', but uranium buyers have been critical of the vagueness of the 'world market price' concept.[24] There has also been widespread dissatisfaction with the ratchet effect resulting from the combination of 'base' and 'world market price'. If the 'world market price' rises, producers reap the full benefit. If it falls, producers are protected by a high and rising 'base' price. Consumers feel that such arrangements give a one-sided advantage to producers. Base prices in market price contracts in the US were in the range of $23-46 for 1978 deliveries, and $37-68 for 1985 deliveries. Actual settlements of market price contracts in the US averaged $41.50 in 1977, while for 1978 an average of $43.95 was reported in January, and $43.65 in July.[25] Although no corresponding non-US data are available, our earlier discussion suggests that the clauses concerning market prices and the price levels in contracts with European and Japanese buyers were not very different from the above.

It may be useful at this stage to summarise our findings with regard to uranium prices. Our survey has documented that uranium prices in major markets outside the socialist countries rose about sixfold in the three-year period from mid-1973 till mid-1976. This rise in prices affected spot deals and new long-term contracts in equal measure. Though the low prices of many old contracts have kept average delivery prices below current market levels in recent years, many producers succeeded in renegotiating their old agreements at substantially higher price levels. Thus, the explosive price development in the mid-1970s has had a widespread impact on the uranium market.

Nominal prices continued to increase from late 1976 until mid-1978, though at a far more moderate rate than in the three preceding years. This is reflected in the NUEXCO spot quotations appearing in Chart 1.1. The 1976 annual average was $39.70. It had reached $42.20 in 1977, and $43.20 in 1978. The 1979 average was $42.60. The price rise as recorded by our evidence has been not only explosive, but also lasting.

The period of the 1970s has been characterised by a very rapid rate of international inflation On top of this, the US dollar has been depreciating against other major currencies. In scrutinising the price series presented above, as well as cost developments to be discussed later in the book, we require an appropriate price index, by which the nominal dollar figures can be deflated. A variety of price indexes are widely used as deflators. The most common ones are probably (a) the US wholesale price index, (b) the US dollar GDP deflator for all OECD countries, and (c) the index of US dollar prices of manufactured goods exported from eleven major industrialised nations. Taking 1968 = 100, these three indexes had reached 189, 207 and 248 respectively by 1977.

It may be interesting to compare these with a specialised index of relevance in our endeavour, namely the Marshall and Swift Index of capital equipment cost, mining and milling in the US. Between 1968 and 1977 this index rose by 89 per cent, i.e. exactly as much as the general wholesale price index in the US. Given that our investigation is international in nature, heavily centred on the industrialised countries, but not particularly related to manufactures exports, we have found the GDP price deflator in dollar terms for the entire OECD area the most appropriate index for our purpose. This index gives a weighted average of the dollar's purchasing power in different OECD countries for all goods and services which enter the GDPs of these countries. We shall be using this price index throughout the book when converting current dollars into constant dollars.

Table 2.3 presents the US dollar general GDP deflator for the OECD area for the period between 1965 and 1979, with 1975/6, the mid-point of the 1970s, equal to 100. The right-hand column in Table 2.3 has been added for reference. Prices and/or costs in real 1975/6 dollars can be obtained by multiplying current dollar values of any of the years shown by the factor appearing in the column for that particular year.

Table 2.3: US Dollar GDP Deflator for the Entire OECD Area

Year	Index 1975/6 = 100	To Convert Current $ into 1975/6 $, Multiply by
1965	49	2.04
1966	50	1.99
1967	52	1.92
1968	53	1.88
1969	55	1.80
1970	59	1.69
1971	63	1.58
1972	70	1.44
1973	79	1.26
1974	87	1.15
1975	99	1.01
1976	101	0.99
1977	110	0.91
1978	127	0.79
1979	139	0.72

Note: The 1979 figure is the author's own forecast, implying a rate of inflation of 9.5 per cent between 1978 and 1979.

Source: Elaborated from data in *Commodity Trade and Price Trends* (1979 edition), World Bank Report No. EC–166/79.

In Chart 1.1 in the introductory chapter, we presented both the current $ NUEXCO spot quotation, and the same series in constant 1975/6 dollars, obtained by the use of the above index. As noted before, the nominal uranium quotation in mid-1976 was six times as high as in mid-1973. In real terms, the mid-1976 price was only 4.9 times as high as three years earlier. After an exceedingly fast rise in the preceding years, the real spot price experienced significant declines from the latter half of 1977, with the end-1979 quotation more than 25 per cent below that of June 1977. Even after taking the falls in real spot prices of recent years into account, the average 1979 quotation was almost four times as high, in real terms, as the average quotation for 1973. Furthermore, the fall in spot prices discussed here is of some-what limited significance. As indicated above, the dominant proportion of uranium sales takes place under long-term contracts. In most of these, producers have protected themselves against inflation for many years to come, through liberal escalation clauses. So far at least, there have been no reports of successful efforts by buyers to renegotiate old contract prices downwards in line with the falling spot quotation.

The set of figures in Table 2.4 provides a simple summary of the development of nominal and real annual average spot prices for uran-ium, taking the 1973 NUEXCO spot quotation as equal to 100.

Table 2.4: Nominal and Real Annual Average Spot Prices for Uranium, 1973-9

	Nominal	Real
1973	100	100
1974	171	159
1975	369	298
1976	618	487
1977	657	477
1978	672	423
1979	666	381

A Comparison with Other Commodity Markets

Our price analysis so far has concerned itself with uranium prices in isolation. It provides in our view a sufficient background for a compari-son of uranium price developments with the price experience of a wide array of other internationally traded commodities. Such a comparison is required to establish whether price movements like those of uranium are common or not in international commodity markets. The exercise conducted below uses data compiled by the World Bank. The coverage comprises 32 major internationally traded commodities for the period 1920-55 and 46 commodities for the period from 1955 onwards, and provides annual average price data for each in nominal as well as real terms.[26]

In our analysis we intend to compare the upward price change in uranium in the mid-1970s with upward price movements in other commodity markets since 1920. The use of annual data assures that very short-run and hence less significant price peaks are disregarded. Our comparisons are made both in nominal and in real terms. Apart from studying the relative magnitudes of the upward price changes, they also consider the abruptness in the price rise and the perseverance of the high price levels. The last two conditions are included in our comparison because, as discussed earlier in this chapter, we feel that they provide additional dimensions to the significance of price changes.

Noting that uranium prices rose more than sixfold in nominal terms between 1973 and 1976, and almost fivefold in real terms, we have searched the World Bank data for cases where individual commodity prices experienced at least a quintupling in nominal prices *or* a tripling in real prices. The exercise would not be very meaningful if we did not set a limit on the period of the price rise. Certainly, in nominal terms at least, most commodity prices have experienced fivefold rises in the period from the late 1940s to the late 1970s. A significant feature in the uranium market was that the price change occurred within a time span of only three years. In our search for similar price experiences in other commodity markets we set a less strict time limit: the nominal *or* real price increases stipulated above should have occurred within a time span of seven years.

Very few products pass the above criteria in the 1920-55 period. In the material studied we have in fact found only two clear-cut instances. One is rubber (London quotation), whose real price rose between 1921 and 1925 by a factor of 4.2. The other is Portuguese West African coffee, whose New York quotation increased between 1948 and 1954 by a factor of 3.2 in real terms. In both cases, the boom was less pronounced and much more short-lived than in uranium in the present decade. Both rubber and coffee prices declined sharply, in nominal as well as real terms, in the years following the peak.

In two other cases, the real quotation increased by a factor of 2.9 –

rubber between 1949 and 1951 and maize between 1948 and 1951 (both quoted in London). Again, the high price levels endured for a short time only, and prices declined rapidly after 1951.

In the 1955-78 period, only 6 of the 46 products studied by the World Bank passed the above criteria. These were:[27]

cocoa (International Cocoa Organization, average daily price of nearest three trading months);

sugar (International Sugar Organization, London daily price in bulk, c.i.f. UK);

coal (bituminous, export unit value, f.o.b. US ports);

petroleum (Saudi Arabian, average realised price, f.o.b. Ras Tanura);

phosphate rock (f.a.s. Casablanca);

triple superphosphate (f.o.b. US Gulf ports).

The price histories of these products, along with that of uranium during the upswing phases, are reproduced in Charts 2.1 and 2.2. The charts illustrate the price developments in the form of indexes, where the year when the price rise started is given as 100. This method makes it easy to compare the relative magnitudes of the price rises experienced by each commodity, and it also illustrates the abruptness in the respective price booms. In the upper charts, the price changes are expressed in nominal dollar terms, while the lower ones illustrate the price increases in real terms.

Of the six commodities listed, only sugar and petroleum have recorded higher nominal price increases than uranium. In real terms, the uranium price record is surpassed only by that of sugar. Four of the commodities – sugar, cocoa, phosphate rock·and triple superphosphate – can be distinguished from uranium and petroleum by the relative brevity of their price booms. After reaching their peaks, the prices of the four have fallen sharply in the following year, providing relief for consumers. In contrast, the prices of uranium and petroleum have remained at the high level for over three years.

Earlier in this chapter we discussed the marginality of the world sugar market. The extreme peak of sugar prices should be seen in that context. The marginality may even provide a major explanation of the volatility in that price. Though sugar prices swung higher than uranium prices, the brevity of the peak and the marginal nature of the market substantially reduce the significance of the sugar price boom.

Only petroleum has experienced a price development similar to that

Chart 2.1: Commodity Price Increases over the Boom of the 1970s

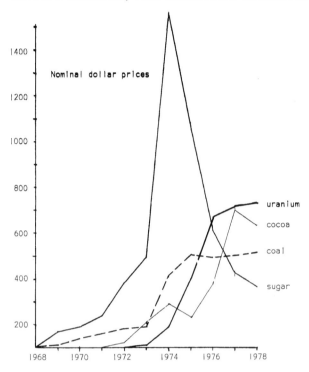

Index, year when price increase started = 100.
See text for explanation.

Sources: NUEXCO for uranium and *Commodity Trade and Price Trends*, 1979
edition, World Bank Report No. EC 166/79, for all other commodities.

Chart 2.2: Commodity Price Increases over the Boom of the 1970s

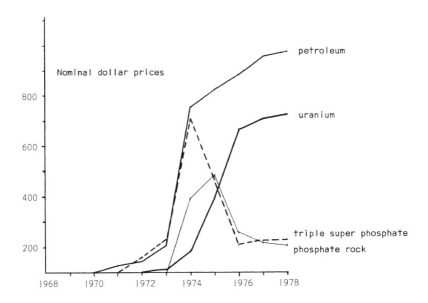

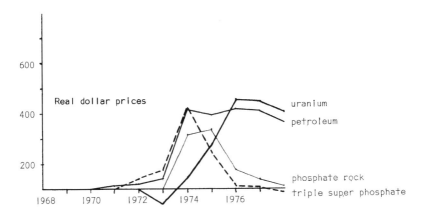

Index, year when price increase started = 100.

of uranium, if sharpness, size, breadth of impact and perseverance are taken into account. Although the petroleum price rose more abruptly, the magnitude of its increase from the start of the price boom through the 1976-8 period was dwarfed by that of uranium. In 1979, however, petroleum prices rose substantially both in nominal and in real terms, all while the nominal uranium price remained stagnant and the real uranium quotation declined.

Our exploration of price booms in international commodity markets over the past sixty years clearly demonstrates that the developments in uranium in the mid-1970s have been exceptional. Apart from petroleum, there are hardly any precedents in terms of abruptness, generality and magnitude of the price rise, and the subsequent durability of the high price levels, comparable with those experienced in the uranium market.

The drastic developments in the uranium market are surprising, given the considerable foreseeability of both requirements and supply afforded by the long lead times between decisions to invest in new uranium mines or in the construction of nuclear reactors on the one hand, and the completion of such investments on the other. The exceptional performance of uranium prices provides an important rationale for a study exploring the causes of this performance. The rest of the book is devoted to this task.

Notes

1. See, for instance, M. Radetzki, 'Market Structure and Bargaining Power, a Study of Three International Minerals Markets', *Resources Policy* (June 1978).

2. Euratom Versorgungsagentur, *Annual Report 1971*.

3. *Uranium Price Formation*, EPRI (October 1977), prepared by Charles Rivers, Cambridge, Mass.

4. G.F. Combs and J.A. Patterson, 'Uranium Market Activity', paper presented at Uranium Industry Seminar, Grand Junction, Colorado, October 1978.

5. More precisely, the 'exchange values' represent NUEXCO's judgement of the lower range of sellers' offers until December 1973, and of the higher range of buyers' bids since that time.

6. *Uranium Price Formation*, EPRI (October 1977), pp. 3-35. See also Chapter 3.

7. *Mineral Facts and Problems, 1975 Edition*, US Bureau of Mines, Bulletin 667 (Washington, DC, 1976).

8. Euratom Versorgungsagentur, *Annual Report 1974*.

9. Euratom Versorgungsagentur, *Annual Report 1975*. The difference between US and European prices is difficult to determine, since Euratom does not give the dates when the quoted price level occurred. Spot prices in the US were about $26 in September 1975.

10. Euratom Versorgungsagentur, *Annual Report 1976*.

11. *Nukem Market Review*, no. 3 (1976).

12. *Metals Analysis and Outlook*, no. 4 (September 1977), Charter Consolidated, London.

13. *Nucleonics Week*, 23 June 1977.

14. *Nukem Market Review, Nuclear Fuel* and *NUEXCO Market Report*, various issues. Also Euratom Versorgungsagentur, *Annual Report 1978*.

15. For the experience in the US, see, for instance, an internal letter by H.M. Weed in the Anaconda Company, of 14 June 1974, presented as evidence with ref. no. VII, 4.86 in legal case T 192-194/75 at the District Court of Stockholm, Dept. 7.

16. Euratom Versorgungsagentur, *Annual Report 1976*, and F. Oboussier, 'Statement on the World Needs for Uranium, its Demand and Production Possibilities' (Bonn, September 1977).

17. *Nucleonics Week*, 9 September and 21 October 1976.

18. *Nucleonics Week*, 23 December 1976.

19. Swedish State Power Board Contract with Uranex of 13 June 1973, and SSPB Memo. of 4 May 1977, signed by I. Wivstad.

20. *Nukem Market Review*, no. 2 (1976).

21. *Nucleonics Week*, 12 February 1976.

22. *Nucleonics Week*, 23 December 1976.

23. Combs and Patterson, 'Uranium Market Activity'. For a comprehensive survey of contracting conditions in uranium, see M.M. Pottier, 'Mechanisms of the Uranium Market' in *Uranium Supply and Demand* (The Uranium Institute, Mining Journal Books Ltd, London, 1978).

24. See, for instance, Oboussier, 'Statement on the World Needs for Uranium', and Pottier, 'Mechanisms of the Uranium Market'.

25. Combs and Patterson, 'Uranium Market Activity'.

26. *Commodity Price Trends*, 1970 edition, World Bank Report No. EC-166/70, and *Commodity Trade and Price Trends*, 1979 edition, World Bank Report No. EC-166/79. Varied price deflators are used to obtain the real price series in the two documents. While these deflators are not strictly comparable with the one we use to obtain real uranium prices, we feel that the differences are not large enough to affect in any significant way the conclusions which follow from our analysis.

27. Where several quotations for a commodity are given, only the one with the sharpest price rise has been included.

3 THE MILITARY LEGACY

Introduction

The commercial uranium market, the focus of attention in the present book, did not become properly established until the late 1960s. A full understanding of its functioning is impossible without an analysis of the preceding phase in which uranium demand was dominated by nuclear defence programmes. The following pages are intended to provide the appropriate background. In a sense, the present chapter forms a digression, but a very important one, from our subject proper. After having dealt with the implications of the large but short-lived military needs for uranium in the 1950s and early 1960s, we resume our main theme in Chapter 4.

Production of uranium was initiated on a significant scale in the late 1940s in response to the urgent military requirements that developed at that time. Output rose very fast in the following decade, to reach a maximum of 44,000 short tons in 1959. This level, which was more than ten times the output of 1949, was not attained again until 1978.

Around 1960, when military requirements became saturated, the abrupt reduction in demand left the uranium industry painfully under-utilised. From 1959 to 1965, production fell by more than half, as the industry tried to adjust to the changed conditions. Demand was reduced by even more. The excess output was absorbed with the help of governmental and industry inventory schemes aimed at keeping the industry alive until the anticipated emergence of large-scale demand from nuclear reactors in commercial use.

Commercial demand was slow in developing, however. In 1970, the need for uranium for running existing reactors was only 12,000 short tons,[1] compared to production of more than 24,000 short tons. As recently as 1977, uranium requirements remained at below 30,000 short tons. Production in that year, at 36,000 short tons, was still one-third below the peak production attained in 1959.

Most of the 1960s constituted a difficult period for the uranium industry. The problems that were experienced clearly sowed the seeds for future world-wide cartel collaboration among producers. The non-US uranium industry was more severely hit by the abrupt shrinkage in defence needs. The military dominance of the US through the 1950s, and the consequent virtual American monopoly in several phases of

Chart 3.1: Western World Uranium Production, 1949-78

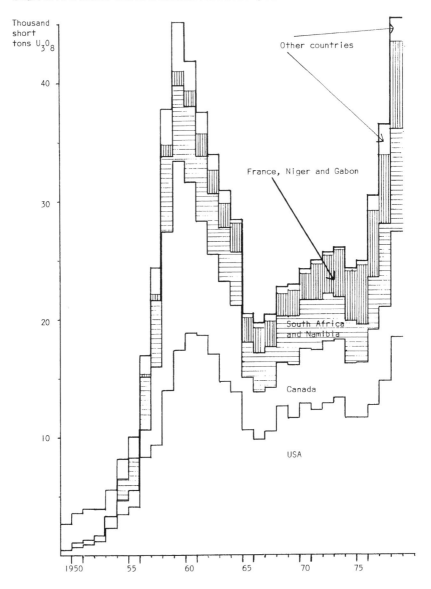

Sources: *US Mineral Resources Geological Survey*, Professional Paper 820, (Washington, DC, 1973); OECD, *Uranium Resources, Production and Demand* (OECD, Paris, 1969, 1973 and 1977); *Mining Annual Review* (1978 and 1979); *Applied Atomics* (31 July 1979).

nuclear technology, permitted US policy-makers to discriminate in favour of domestic uranium producers.

Incentives and Production Growth

Through the late 1940s intensive discussions were taking place among the allied powers' governments about the establishment of an arsenal of nuclear weapons. The US, UK and Canada were the leading partners in these discussions. A major issue in the deliberations was how and where to procure the necessary uranium supplies to carry out the weapon programmes.

Broad agreements were reached in 1948, with the formation of the Combined Development Agency, set up jointly by the US and UK governments to assure adequate uranium supplies.[2] The Belgian Congo (now Zaire) accounted for a major proportion of the limited world production at that time. Most of the supplies from this country were allocated to the US, where needs were greatest and most urgent. But since the military endeavours required a manyfold production increase, the Agency initiated a very broad-based programme of exploration incentives, financial support for investments and price guarantees, in order to encourage a fast expansion of production.

The effort, which centred on the US, Canada, South Africa and, to a limited extent, Australia, was highly successful. The intensive exploration activity resulted in a fast build-up of mineral reserves. For instance, US reserves with a forward cost of $8[3] went up from an insignificant figure in 1952 to more than 100,000 short tons by 1956 and to close to 200,000 short tons in 1959.[4] Given the considerable lags between investment decision and production start-up, the large increases in output came only after 1955. In the US, production went up from 1,200 short tons in 1952 to 4,100 short tons in 1955 and peaked at 18,800 short tons in 1960. In Canada, the 1,000 short ton level was reached in 1953/4, and output expanded speedily in the following years to reach almost 16,000 short tons in 1959. Production in South Africa went up from 1,800 short tons in 1954 to almost 6,500 short tons in 1959. Australia also contributed to the Combined Development Agency efforts, though on a smaller scale. Production in 1958 was 600 short tons. Between 1959 and 1963, average annual output remained between 1,000 and 1,500 short tons. In contrast to these experiences, production in the Belgian Congo declined sharply in 1960 and was discontinued in the following year, as the country's uranium reserves were depleted. France, which entered the nuclear arms race somewhat later and somewhat independently of the US/UK efforts, developed domestic supplies of uranium only in the mid-1950s. These were supplemented by production under considerable French control in Madagascar, and later in Gabon and Niger.

The US government dominated the efforts to expand uranium production. Through its Atomic Energy Commission (AEC), the US acquired roughly 85 per cent of the non-socialist world output of uranium up to 1960s, while most of the balance went to the UK.[5] The expansion of world production can be explained, by and large, by the favourable conditions under which the AEC procured its uranium requirements. Encouragement to uranium-mining in the US was provided directly through the AEC. In other countries the activities were directed under the auspices of the Combined Development Agency.[6]

The price history for the military purchases is incomplete. The AEC bought uranium under a variety of conditions, i.e. at fixed prices over the whole period of a several-year contract, at prices which escalated annually, or at cost-plus formulas. Table 3.1 summarises the annual average prices paid by the AEC to US producers from 1950 until the programme was discontinued. It appears from the table that prices, both in nominal and in real terms, rose from 1950 to reach a peak in 1953, and declined almost without interruption thereafter, with the real price obviously falling much more than the nominal one.

At least in the period 1957-63, prices paid by the AEC to non-US producers were higher than those paid domestically. Thus, the prices received by Canadian producers exceeded the ones paid to US producers by 1.3 per cent in 1957, 24.7 per cent in 1960 and 23.7 per cent in 1963. Supplies from other countries, dominated by South Africa, were priced even higher, with average prices 10.7 per cent, 35.7 per cent and 44.4 per cent above those paid to US producers in the three years.[7] The reasons why the US authorities paid more for foreign supplies have never been clearly stated.

There are little data about prices received for deliveries to countries other than the US. Canadian and South African producers were receiving about the same prices for their deliveries of uranium to the US and UK in the 1950s, with the South African price about $1/lb above the Canadian one.[8]

Quite clearly, the extremely fast expansion of the uranium industry, raising output almost fivefold within the short time span between 1954/5 and 1959/60, was mainly in response to the very favourable prices offered by the US AEC from the late 1940s onwards. We may note that the average prices between 1950 and 1955, of between $30 and $35 (fixed 1975/6 $), offered to US and non-US producers were considered sufficiently attractive by the industry to cause a scramble by miners to increase output. We will return to this point in Chapter 4.

Table 3.1: Prices Paid by US Atomic Energy Commission for Uranium Purchases from Domestic Producers

Fiscal Year	Annual Average Prices per lb U_3O_8	
	Current $	Constant 1975/6 $
1950	9.11	30.19
1951	10.10	30.48
1952	11.28	32.31
1953	12.35	35.38
1954	12.27	34.56
1955	12.25	33.94
1956	11.51	30.88
1957	10.49	26.46
1958	9.45	23.15
1959	9.12	22.02
1960	8.75	20.54
1961	8.50	19.41
1962	8.15	18.12
1963	7.82	16.94
1964	8.00	16.90
1965	8.00	16.29
1966	8.00	15.90
1967	8.00	15.36
1968	8.00	15.02
1969	6.99	12.57
1970	5.74	9.70
1971	5.54	8.75

Notes: The constant price series is obtained by deflating nominal prices by the US dollar general GDP deflator for the entire OECD area, as given in *Commodity Trade and Price Trends*, 1979 edition, World Bank Report No. EC 166/1979.

Until 1976, the US fiscal year ended 30 June of the year stated. From then on, the US fiscal year ends on 31 October of the year stated.

Source: USAEC, GJO — 100, January 1976, p. 11.

The positive and strong producer response enabled the authorities in the US and UK to satisfy their requirements of uranium for the nuclear weapons programmes in a relatively short time. The US AEC purchases for this purpose, from domestic and foreign producers, are depicted in Chart 3.2. The total rose roughly in parallel with world production, to reach a peak in 1959. Until 1960, more than half of the total procurements came from non-US producers. From 1961 onwards, US supplies played a major and increasing role in satisfying remaining US needs.

Chart 3.2: US AEC Purchases of Uranium, Fiscal Years 1949-71

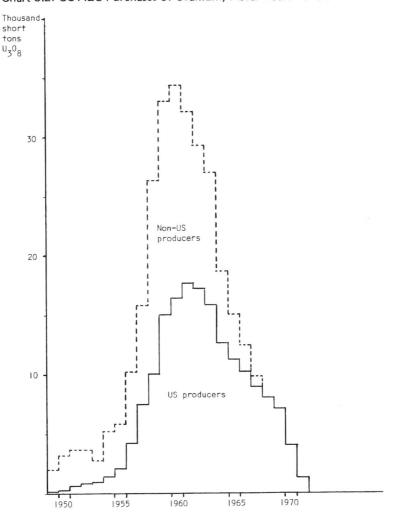

Source: *Uranium Price Formation*, EPRI (October 1977), prepared by Charles Rivers, Cambridge, Mass.

It should be clear that the reduction in production and procurement of uranium were not caused by constrained supply. On the contrary, the sharp turnaround in 1959/60 came about because the producer response to the incentive prices had been more favourable than expected and had satisfied the military needs faster than anticipated. The signs of excessive supply emerged as early as 1956. In 1957 and 1958 the US AEC began to restrain the growth of the industry by modifying the contract terms for future deliveries.[9] After 1959, procurements of foreign uranium were sharply curtailed, and in 1967 they were completely discontinued. Procurements from domestic suppliers reached their peak in 1961. In order to help the US uranium industry to survive, the AEC instituted a variety of stretch-out arrangements so that contracted quantities could be delivered over a longer period of time. AEC procurements were discontinued at the end of 1970. Soon afterwards the Agency declared that it had a surplus stock of uranium amounting to some 50,000 short tons U_3O_8 which it planned to dispose of by seeking competitive bids.[10] The UK Atomic Energy Agency, the second buyer of uranium, was reported to have been overstocked as early as 1961.[11] Hence, the industry could not expect any boost in uranium demand from that source, as the US procurements declined.

The Consequences of Saturated Military Requirements

In the next chapter we will track the developments of the commercial uranium market in considerable detail. At this stage, it suffices to note that commercial demand remained limited throughout the 1960s, and could certainly not replace the reductions in military procurement. This created serious difficulties for producers. The AEC's threat to dispose of its surplus stock was an additional depressing factor in uranium trade negotiations. The industry was forced to accept continuously decreasing prices for the remaining deliveries to the military authorities. The AEC price developments appear in Table 3.1. In 1962, after lengthy negotiations, the UK Atomic Energy Agency was induced to sign a new contract for Canadian uranium deliveries until 1970. But given that the UK was already well supplied, and that the Canadian industry lacked alternative sales outlets, the average price agreed upon for this contract was $5.03 per lb,[12] substantially below the prices paid by the AEC through the 1960s.

Reduction of prices was of very limited help in increasing demand, however. Alternative uses for uranium, at a sufficient scale to keep the industry busy, simply did not exist at that time. Capacity utilisation had to be sharply curtailed in the latter half of the 1960s. In 1967, for instance, annual world capacity was assessed at more than 35,000 short tons,[13] but output was barely above 20,000 short tons. Numerous producers left the business altogether during these years.

The uranium industry was differently affected in each producing country. In the US, as already noted, the AEC concentrated its procurements after 1963 on uranium mined within the country. This reduced somewhat the problems experienced by US producers. The prolonged stretch-out period in AEC's procurements also helped the industry in its downward adjustment process. By 1970, the last year of significant AEC purchases, the miners succeeded in selling more than 9,000 short tons to commercial buyers in the US. Foreign producers were barred from competition in this market by a US government decision that foreign uranium could be enriched in the US only if it were subsequently re-exported.[14] In this way the US administration used its monopoly in the field of enrichment for the purpose of commercial protection of the country's uranium producers.

In Australia, where annual production never exceeded 1,500 short tons, the industry was not of great national importance, and could be dismantled without serious pains. Production declined from 1,200 short tons in 1963 to insignificant amounts in the 1964-70 period. In the absence of export demand, the output of these years was stockpiled by the Australian Atomic Energy Commission.[15] In 1970, production was discontinued altogether.

South African uranium production during the 1960s was in its entirety a by-product of gold, and most mines were viable on the basis of gold alone. In 1959, when output peaked, there were 26 uranium-producing gold-mines, supplying more than 6,000 short tons U_3O_8 per year to the Combined Development Agency. With the decline in military purchases, many of these mines stopped uranium production promptly. By 1965, 19 of the former producers had discontinued uranium activities altogether, and production fell to less than one-half the earlier level.[16] Dumps for uranium containing gold tailings were carefully kept by the mines, so as to facilitate extraction, if and when market conditions improved.

Canada was probably most severely hurt by the sudden shrinkage in world uranium demand, because it was the producing country in which uranium weighed most heavily in the national economy. At its peak in 1959, Canadian output reached 15,900 short tons produced in 23 different mines, valued in excess of Can \$310 million.[17] In that year, uranium was the country's fourth-largest commodity export, representing close to 6 per cent of the overall export value. As demand declined, production was curtailed. In 1967 only four mines remained in operation, with an output below 4,000 short tons, valued at less than Can $50 million.[18] In the face of declining markets, and to prevent a total erosion of its uranium industry, the government instituted two successive stockpiling programmes in 1963 and 1965 respectively, with financial requirements in excess of Can $100 million. Producers also stockpiled uranium to maintain a nucleus of a uranium industry in the country.[19]

The government of Canada reproached the US administration for decisions taken in 1964-6 prohibiting the enrichment of foreign uranium for domestic use. In a memo, dated August 1969, a spokesman for the government pointed out that the Canadian industry had originally been developed to satisfy US needs, that current production in Canada was only 25 per cent of peak production of ten years earlier, and that the US moves effectively barred the Canadian industry from competing in the US commercial market which at that time represented more than 50 per cent of the world market.[20]

The importance of uranium to the Canadian economy, the collapse of international uranium demand and the unresponsive attitude of the US administration to the Canadian problems prompted the government of Canada to take the initiative in the early 1970s towards the establishment of international marketing arrangements for uranium, aimed at preventing excessive supplies and assuring reasonable prices to producers. Representatives of governments and private producers from Australia, Canada, France and South Africa took part. Some US producers also participated in the arrangements. The Canadian initiative, originally taken to help the country's industry to survive, constituted the origins of the international uranium cartel of the 1970s. We will return at length to the cartel subject in Chapter 7.

The French uranium industry and its affiliate production units in African countries led their lives a bit apart from the developments outlined above. As already noted, France's nuclear weapons build-up started later than in the US and UK. The French government was itself profoundly involved in uranium production, and its needs continued to grow. Uranium production in France and its African affiliates did not experience a decline in the 1960s, but has continued to grow steadily up to the present.

With the exception of France, however, the world uranium industry enetered the commercial era in the late 1960s in a state of depression.

The curtailment of military requirements had forced numerous producers out of business. Current production was far below the huge capacity created by the military incentive programmes. The surplus stocks held by government agencies and private producers overhung the market and depressed prices further. The immediate future looked bleak indeed, especially when compared with the highly profitable conditions of the past. The industry had a strong incentive to survive, however. Projections into the mid-1970s and beyond pointed to a very strong expansion of demand, as the civil nuclear-power generation programmes of different countries got properly under way.

Notes

1. OECD, *Uranium Resources, Production and Demand* (OECD, Paris, September 1970).

2. R.G. Hewlett and F. Duncan, *Atomic Shield, 1947/1952* (Penn State University Press, University Park, 1969).

3. See Chapter 4 for a discussion of reserves and forward costs.

4. *Uranium Price Formation*, EPRI (October 1977), prepared by Charles Rivers, Cambridge, Mass.

5. P. Mullenbach, *Civil Nuclear Power* (Twentieth Century Fund, New York, 1963).

6. Some of the programmes of encouragement outside the US are briefly described for Canada in 'Canada's Uranium Industry Recovers', *Nuclear Engineering* (June 1968), p. 505, or in J.W. Griffith, *The Uranium Industry, its History, Technology and Prospects* (Dept. of Energy and Mines, Ottawa, 1967), pp. 8-22; for Australia in 'The Australian Uranium Industry', *Atomic Energy* (April 1976), p. 20; and for South Africa in R.E. Worrol, 'The Pattern of Uranium Production in South Africa', *Uranium Supply and Demand* (The Uranium Institute, London, 1976), p. 21.

7. Mullenbach, *Civil Nuclear Power*. The price differentials for 1963 were in part estimated.

8. Griffith, *The Uranium Industry*.

9. *Uranium Price Formation*, p. 3.2.

10. Government of Canada, 'Policy Statements and Press Releases on Uranium Marketing', Note No. 359 of December 1971 (Ottawa, undated). See also *Uranium Price Formation*, p. 3.34

11. Griffith, *The Uranium Industry*.

12. Griffith, *The Uranium Industry*.

13. OECD, *Uranium Production and Short-term Demand* (OECD, Paris, January 1969).

14. *Uranium Price Formation*, p. 3. 35.

15. A.J. Grey, 'Australian Uranium, Will it Ever Become Available?', *Uranium Supply and Demand* (1976).

16. R.E. Worrol, 'The Pattern of Uranium Production in South Africa', *Uranium Supply and Demand* (1976).

17. In 1959, the value of the Canadian dollar was US$1.04. Between 1963 and 1967, its value varied within the narrow range of US$0.92-0.93.

18. 'Canada's Uranium Industry Recovers', *Nuclear Engineering* (June 1968).

19. Government of Canada, 'Policy Statements', background paper of 22 September 1976.

20. Government of Canada, 'Policy Statements'.

4 THE FUNDAMENTAL FACTORS OF DEMAND AND SUPPLY IN THE INTERNATIONAL URANIUM MARKET

Introduction

A major hypothesis of our study, to be substantiated in this and several following chapters, is that the explosion of uranium prices in 1974 and 1975 was unexpected and came about as a result of the interrelated and simultaneous emergence of several extraordinary factors with a very strong impact on the uranium market. This chapter provides a conventional analysis of major demand and supply forces in the commercial uranium market. The conclusions we reach lend support to the first part of the above hypothesis: from the late 1950s to the mid-1970s there was nothing in the ordinary determinants of uranium demand and supply to suggest a price rise of anything like the magnitude that actually occurred. Neither the market itself nor informed opinion within and without the nuclear industry could foresee what actually happened.

To explain satisfactorily the uranium price behaviour, we have to venture beyond the uranium market proper, and analyse the set of extraordinary circumstances which emerged in the 1973-6 period, and which together exerted a profound influence on uranium prices. This is the task to be tackled in the next few chapters.

The present chapter is organised as follows: Our scrutiny of ordinary demand and supply forces starts out with a brief description of uranium requirements in the nuclear fuel cycle. We then discuss the problems of the industry in the late 1960s and early 1970s due to excess supply and low prices, resulting from the sudden discontinuation of military demand. In this connection, consecutive forecasts of future demand are presented. Special attention is devoted to the ability of the industry to cope with the forecast growth in demand. For this purpose, we analyse in detail the supply prospects for uranium. Actual and anticipated developments in production capacity and in reserves are scrutinised and related to current and future demand. Production costs and the prices required to encourage investments in new capacity are determined as far as data permit. A variety of forecasts of future uranium prices, formulated at different times up to and including 1974, are presented. The chapter ends by noting the paradox of a set of key demand and supply factors which would have suggested the continuation of depressed conditions throughout the period of exploding prices.

47

Chart 4.1: The Nuclear Fuel Cycle

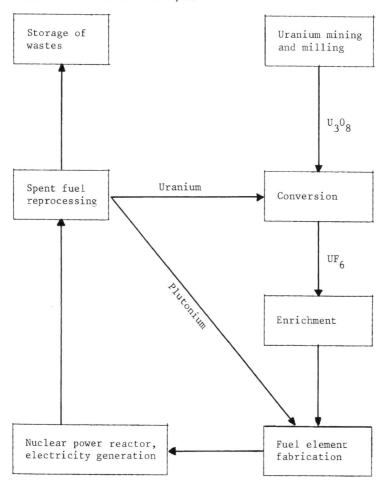

The Elasticities of Uranium Requirements in the Nuclear Fuel Cycle

Nuclear power can be generated through a variety of processes. From the late 1970s and until the end of the present century, the light water reactor is expected to account for about 90 per cent of total nuclear capacity in the Western world. Future demand is therefore regularly assessed on the basis of uranium use in this type of reactor. Chart 4.1, along with the following discussion, tries to clarify briefly the various steps of uranium transformation, and to point to the possibilities of varying the quantity of natural uranium required to obtain a unit of energy.

'Yellowcake' is the term applied to uranium concentrate, the end product of uranium *mining and milling* operations. Quantities of uranium are commonly measured in terms of the content of uranium oxide, U_3O_8, in the concentrate. U_3O_8 regularly constitutes 80-85 per cent of yellowcake. In some countries, the content of U, the pure uranium metal, in yellowcake is used to measure uranium quantities; U_3O_8 contains 84.8 per cent U.[1]

The U_3O_8 concentrate goes through a *conversion* process, resulting in uranium hexafluoride, UF_6, which is gasified to facilitate enrichment.

Natural uranium contains 0.7 per cent of the radioactive isotope U_{235}. For use in the light water reactor, natural uranium has to be enriched, to increase the U_{235} content to about 3 per cent. Uranium *enrichment* is a complex and laborious process, the costs of which are comparable to the cost of producing U_3O_8. The quantity of natural uranium required to obtain a unit of enriched uranium depends on the amount of U_{235} discarded in the enrichment tails. The higher the U_{235} content in the tails (referred to as the tails assay), the larger the amount of natural uranium needed for a given quantity of enriched material. The US authorities, which held a virtual Western world monopoly of enrichment services until the late 1970s, have been supplying their customers with enriched uranium in exchange for natural uranium deliveries on the basis of a tails assay of 0.2 per cent.

The enrichment process is measured in terms of separative works units, SWUs. The lower the tails assay, the more laborious is the enrichment process. The requirement of SWUs rises geometrically for each arithmetic decline in the tails assay. The figures in Table 4.1 indicate the variations in natural uranium and SWUs required to obtain a unit of enriched uranium as the tails assay is changed.[2] The requirements variations are expressed in terms of an index which equals 100 at a tails assay of 0.20 per cent. At this tails assay about 1 lb of natural uranium and 0.3 SWUs are required to obtain 0.15 lb of enriched uranium, containing 3 per cent of U_{235}. The figures indicate that natural uranium requirements will rise by 20 per cent if the tails assay is raised from 0.2 to 0.3 per cent, but that this will result in a 20 per cent saving in enrichment capacity. At each set of prices of natural uranium and enrichment services, one can determine the optimum tails assay, i.e. the one at which the overall cost of enriched uranium is minimised.

Table 4.1: Requirements Variations in Natural Uranium and SWUs to Obtain a Unit of Enriched Uranium

Tails Assay	Uranium Requirement	SWU Requirement
0.15	93	116
0.20	100	100
0.25	109	90
0.30	120	80

To provide the relative orders of magnitude, it may be pointed out that the annual requirements of a light water reactor with a capacity of 1000 MWe amount to about 200 short tons U_3O_8 and 125,000 SWUs.[3]

Enriched UF_6 is the major input in the *fabrication of nuclear fuel elements*, which are then used in nuclear reactors for power generation.

After initial *loading of the nuclear reactor*, between one-third and one-quarter of the reactor fuel core is removed and replaced with new fuel elements on an annual basis. Though only part of the fissionable material has been burnt in the elements which are removed, their efficiency deteriorates due to the accumulation of waste products.

The spent fuel elements can be sent for *reprocessing*, to extract the remaining fissionable material for use in the production of additional nuclear fuel. Uranium extracted in the reprocessing of spent fuel elements is sent for conversion into UF_6. Plutonium created through the fission processes when the fuel is in use is extracted and sent for the fabrication of plutonium fuel elements, which are then employed in the same way as ordinary uranium fuel elements. The remains of the spent fuel are sent for waste storage.

Reprocessed uranium can provide about 19 per cent of the overall fuel required in a reactor after initial loading. Plutonium extracted in reprocessing can provide for an additional 11 per cent of the overall fuel need.[4] Full reprocessing can therefore reduce the overall need for natural uranium by 30 per cent.

The Combination of Low Prices and Fast Requirements Growth Forecasts

As already noted in Chapter 3, the excess capacity resulting from the reduction of military requirements before a sufficient alternative demand had developed to compensate for the shortfall led to an extended period of depressed prices in the commercial uranium market. Interrupted by a brief improvement, with strengthened prices in late 1968/early 1969, this period lasted until 1973. The price history of uranium over this and the subsequent period has been discussed in detail in Chapter 2.

It is clear that numerous uranium producers were unable to cover their full costs, including a reasonable return on invested capital, from the prices that prevailed in the late 1960s and early 1970s. Mining firms reacted in a variety of ways to the depressed demand for their product. A large number left the business altogether. Among these were the high-cost ventures lured into production by the attractive military conditions, the firms with a weak financial structure that could not stand up against the strain of extended price depression, and those pessimistic about the future prospects for demand. The hardship for those who remained in business was mitigated by these exits. When evaluating the industry's difficulties in this period, one must also consider the fact that prices during the military stage had been set at levels permitting premature depreciation of capital assets in many cases. In any case, given the high capital-intensity in the uranium industry, operating costs were low, and the prices received were sufficient to provide for their coverage. Production was therefore continued, but in view of the low return on capital, re-investments were held back, and the production capability of existing installations dwindled over time. The prices that prevailed were in themselves insufficient to induce investments in additional reserves or new mining and milling capacity in most cases.

Yet, some such investments were taking place. In 1972, Continental Oil, Standard Oil (NJ), Sohio and Gulf Oil in the US were reported to have entered, or be considering entry, into uranium production.[5] Prospecting for new deposits continued. In the US, for instance, exploration and development drilling for uranium reached a peak of almost 30 million feet in 1969, more than three times the previous high which had occurred in 1957.[6] In Namibia, development work on the large-scale Rossing deposit had been initiated in the late 1960s, and continued through the early 1970s.[7] In Australia, production had ceased in 1964, but from 1970 onwards the search for uranium was reported to have increased significantly, leading to several important finds.[8] In some of the discoveries, overall production costs appeared to be so low that production would be profitable even at the then prevailing low price levels.[9]

The major reason for maintaining and in some cases even expanding capacity during this period appears to have been not the current price level as such, but the prospects of a very fast rise in demand, and the expectation of a consequent improvement in prices.

The optimism about future expansion in nuclear electricity-generating capacity increased considerably during the 1960s, to culminate around 1970. Table 4.2 provides the details of forecasts made at different points in time. It also presents the consequent forecasts of uranium requirements to feed the nuclear power plants. Developments in the two series are not proportional, *first* because the time sequences of expansion differ between the forecasts, and larger amounts of uranium are needed for the first reactor loading than for subsequent ones, and *second*, because the forecasts are based on varied assumptions regarding enrichment tails assays, reprocessing of spent uranium fuel and use of plutonium.[10]

Table 4.2: Forecasts for Installed Nuclear Capacity and Consequent Uranium Requirements (Western World)

Forecast Source and Year	1970	1975	1980	1985	1990	2000
Installed Nuclear Capacity, GWe						
Euratom 1963	17		117			
OECD 1969	26	112	281			
OECD 1970	18	148	300	610		
OECD 1973		94	264	567	1068	
OECD 1975 (high estimate)		69	194	530	1004	2480
OECD 1977 (most probable estimate)			146	278	504	1000
Uranium Requirements, Thousand Short Tons U_3O_8						
Euratom 1963	8	38				
OECD 1969	19	52	90			
OECD 1970	12	37	73	130		
OECD 1973		33	79	140	225	
OECD 1975 (high estimate)		23	69	131	218	407
OECD 1977 (most probable estimate)			53	92	133	231

It may be instructive to take a snapshot picture of the uranium market situation and future prospects as they appeared in 1970. This is summarised in Chart 4.2. As already noted, the state of the industry in 1970 was characterised by an extreme over-supply, with total requirements amounting to 12,000 short tons, as compared to production of 24,000 short tons and capacity around 33,000 short tons.[11] Simultaneously, the envisaged increase in requirements was explosive. Forecast growth rates amounted to 20 per cent per year in the ten-year period 1970-80, and 17 per cent per year in the fifteen-year period 1970-85. Although for some years to come the high levels of existing capacity would ensure adequate supplies, it was clear that very large additions to capacity would be needed from the mid-1970s if the requirements forecasts materialised.

The situation had not changed much by 1973. Capacity had been expanded to 35,000 short tons, and remained far above production at 26,000 short tons. But requirements had failed to grow according to the 1970 forecast. With current reactor needs in 1973 at 22,000 short tons, the extreme over-capacity had shrunk somewhat, but was still large. The 1973 forecast of growth in future requirements, starting off from the lower requirement level actually attained in the current year, had been reduced. It amounted to 18 per cent per year in the ten-year period 1973-83, and to 15 per cent per year in the 15-year period 1973-88.[12]

Even after this reduction, the forecast growth in requirements constituted a strong attraction for producers to stay in business, despite the unremunerative current prices. The forecast expansion in uranium needs was in fact so sharp that questions were raised whether the industry would be in a position to satisfy them. Two issues caused some concern. First, additional capacity might not be established unless the price rose so as to make such investments remunerative. Second, there was a limit set by the amount of existing uranium reserves, beyond which production could not easily be raised. This is because the speed of exploitation of an ore body is constrained by several factors. For instance, the shape of the ore body may constrain the construction of access roads. Because of its physical characteristics, the large Elliot Lake deposit in Canada places limits on the rate of extraction, such that the present reserves could not be exhausted until well into the next century. Furthermore, there is the issue of longevity of mining equipment. If, for instance, the durability of mining installations is ten years, it would ordinarily be uneconomical to install a production capacity exhausting the mine in lesser time. For the level of low-cost reserves

Chart 4.2: The Uranium Situation in 1970[a]

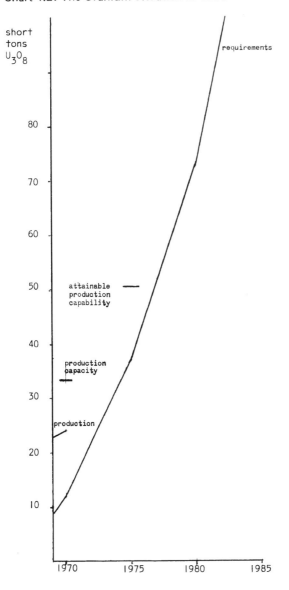

Note: [a] Actual figures for 1969 and 1970, forecasts for 1975-85.

Source: OECD, *Uranium Resources, Production and Demand* (OECD, Paris, September 1970).

existing in 1970, the limit of economic production capacity is shown in Chart 4.2 by the attainable production capability at just above 50,000 short-tons U_3O_8. By 1973, as a result of an expanded reserve base, the attainable capacity figure had increased to 67,000 short tons. Production growth beyond this level would require the use of higher-cost reserves, or an expansion of the low-cost reserve base. Both were thought unlikely to occur unless uranium prices improved. Those who argued for higher prices also stressed the long lead times needed both to identify new reserves and to establish additional production capacity.[13]

We now propose to assess these worries by studying the developments in production capacity and reserves that actually took place, and by analysing the feasibility of continued expansion of production to meet requirements in the longer run. Our analysis is heavily based on the series of OECD reports referred to above. We should point out that the OECD compilations have been readily accepted, quoted and used as reliable summaries of conditions and outlooks at the time of their publication, not only by governments, but also by the uranium industry,[14] as well as by investment advisers and the trade press.

We will first discuss the mine and mill capacity issue. Thereafter we will scrutinise the reserves and resources position, to establish its adequacy and prospects for meeting forecast demand.

Production Capacity Developments

OECD provides figures for the production capability of existing mines and mills at different times. It is instructive to compare these capacities with actual uranium requirements. Since mine and mill construction to exploit existing uranium reserves need not ordinarily take more than three years,[15] we also compare OECD's production capability figures at different times with simultaneous forecasts of uranium requirements four years later, to determine the urgency of the needs to expand capacities. Table 4.3 summarises these data.

On its face value, Table 4.3 suggests that existing production capability was excessive not only up to 1973, but all through 1977. An industry in balance, irrespective of the speed of its growth, should have production capacities continually just above the level of current requirements. In uranium, throughout the period, actual requirements have remained substantially below capacity. Any normal industry assessment would characterise the situation as one of gross and wasteful overcapacity.

Table 4.3: Existing Production Capabilities, Production and Actual and Forecast Requirements (short tons $U_3 O_8$, Western World)

Period of Observation and Forecast	Existing Production Capability		Production		Actual Reactor Requirements		Forecast Reactor Requirements	
	Year	Quantity	Year	Quantity	Year	Quantity	Year	Quantity
1969	1967	38,000	1967	20,000	1967	9,000	1971	25,000
1970	1970	33,000	1970	24,000	1970	12,000	1974	30,000
1973	1973	35,000	1973	26,000	1973	22,000	1977	46,000 (medium)
1975	1975	33,000	1975	25,000	1975	23,000	1979	57,000 (high)
1977	1977	43,000	1977	36,000	1977	30,000	1981	66,000 (high)

Sources: OECD, *Uranium Resources* (OECD, Paris, December 1967); OECD, *Uranium Production and Short-term Demand* (OECD, Paris, January 1969); OECD, *Uranium Resources, Production and Demand* (September 1970); ibid. (August 1973); ibid. (December 1975); ibid. (December 1977).

Only from 1973 onwards did current capacity fall below requirements forecasts four years into the future. This indicates that efforts to expand capacity became warranted from that time. In fact, investments to expand the production capability had been going on at least since 1970. Thus, in 1970, plans were being implemented to expand capacity to 38,000 short tons by 1973. And in 1973, ongoing investments were made with the intention of bringing production capability to 40,000 short tons in 1975. That these capacity levels were in fact not attained[16] was primarily due to the repeated shortfalls in requirements growth.

In reality, mining capacity figures are somewhat ambiguous and elusive. Investments in new mines are required not only for expansion, but also to maintain a stable capacity level, as old mines go out of business due to depleted ore bodies. Considerable uncertainty always surrounds the production potential of moth-balled mining and milling installations. Capacities often refer to the ability of the mines and mills to handle ore rather than to the output of final product. Final-product capacity will increase, for instance, if the miners adopt the practice of high-grading, i.e. of selective extraction of the best parts of a reserve, in order to reduce average costs and maintain required cash flows during low-price periods. This is because with an increase in the grade of ore used, the full capacity mill throughput will give a higher amount of $U_3 O_8$. The obverse may occur as a result of a price increase permitting profitable exploitation of lower-grade ores. Capacity figures often refer to nominal capacities attainable under ideal conditions. In reality, service requirements, breakdowns, strikes, etc., restrict the capacity utilisation. Estimates for the copper industry world-wide over a long period of time suggest average capacity utilisation rates of above 90 per cent.[17] A similar figure ought to be attainable in uranium as well.

Even if the above ambiguities and reservations are taken into account, the impression remains that the mining and milling capacities in uranium through the period studied were far too high in relation to current requirements, and quite adequate, if not excessive, to meet forecast reactor needs.

Could any constraints be perceived on the more long-term capacity expansion to satisfy uranium requirements onward to 1985 or 1990? It is true that the rate of growth in uranium requirements, assessed in 1970 at 17 per cent per year for the 1970-85 period, did look dramatic. However, the corresponding rate of capacity expansion to fulfil requirements in this period was only about 10 per cent per year, because of the excessive capacity in the initial year. A 10 per cent annual growth over a fifteen-year period is not unprecedented in the minerals industries. Western world aluminium production and capacity, for instance, grew by almost 10 per cent per year over the far longer period between 1950 and 1974, to reach a level of more than 12 million short tons in the latter year.[18] The required capacity growth performance can be looked at from yet another angle. Assuming extremely pessimistically that the new uranium-mining and -milling installations would handle very low-grade ores, from which no more than 2 lb U_3O_8 could be extracted per ton of material,[19] the additional capacity to be established between 1970 and 1985 would have to have the ability to handle some 100 million short tons of ore per year. This is certainly impressive, but by no means unattainable. Western world copper industry has added that amount to its handling ability about once every three to four years in the past few decades. Clearly, the problems of keeping uranium production in line with requirements appeared less formidable after 1970, with each consecutive downward adjustment in the forecast expansion of nuclear power capacity.

The Growth of Reserves

Obviously it would be futile to establish mining and milling capacity in the absence of adequate mineral reserves ready for exploitation. We therefore have to clarify the reserves position in uranium in the early 1970s, and study its development over time, to find out whether reserves constituted an impending constraint on uranium supply.

In minerals in general, reserves refer to those reasonably well-defined ore bodies from which minerals can be economically extracted at the time of measurement. This definition leaves considerable scope for ambiguities. Furthermore, the companies which are the owners of reserves are frequently unwilling to publish complete reserve information. Hence, available data usually provide less than full coverage.

Reserves are the result of conscious investment efforts put into exploration. Mining companies will undertake such investments only up to the point needed to assure their forward planning. As existing reserves are used up, new ones are established by exploration to keep the buffer needed by the companies. Increases in price or cost-reducing technological progress will add to the amount of reserves by converting some formerly uneconomic resources into reserves. In mining industries experiencing stable growth, the quantity of reserves ordinarily constitues a relatively stable multiple of current production.[20]

The above interrelationships clarify that it is not very meaningful to compare the current stock of reserves with forecast future mineral requirements for the purpose of deducing when the mineral will be depleted. Reserves provide no indication at all of overall mineral availability. Neither is it possible to use the requirements/reserves relationship for deductions about future price rises due to the need to exploit deposits which have lower ore grades, or which are less economical for other reasons. All experience indicates that additional low-cost reserves will be identified over time. The history of mining also shows that the upward cost push due to the falling metal content in exploited ores has commonly been neutralised by the downward cost push resulting from technological progress.[21]

In uranium, reserves have been defined either in terms of the prices at which the mineral deposits can be remuneratively exploited, or of the costs incurred in extraction. Practices and definitions in these respects vary among countries, and hence the reserve assessments contain considerable ambiguities. Table 4.4 summarises the reserve estimates published by authoritative bodies in different years. Several issues need to be pointed out in relation to this table.

First, we should note that ever since 1970, OECD and others have published reserves in low-cost and high-cost categories. So long as low-cost reserves are ample, exploitation will concentrate on these, and miners will have little interest in exploration for high-cost reserves.

Hence, it should be clear that relatively limited efforts have gone into identification of the high-cost reserve category.

Second, reserves have been classified by OECD in current dollar price/cost categories. In column 3 of the table we have converted the figures into constant 1975/6 dollars. Though not too much significance should be attached to the exact levels of the figures, we may note that the low-cost category remained relatively stable in terms of its constant price/cost limit up to 1975. In the 1977 assessment, the low-cost category limit was raised substantially, while in the 1979 figures it was brought down again to just above the 1968 low-cost category cut-off figure.

Table 4.4: Uranium Reserve Position (Western World, thousand short tons U_3O_8)

Year	Price/Cost Category		Reserve Quantity		Low-cost Reserves in Percent of Forecast Ten-year Forward Cumulative Requirements
	Current $	1975/6 $	Low Cost	High Cost	
1968	below 10	below 19	700		105
1970	below 10	below 17	840		195
	10-15	17-25		750	
1973	below 10	below 13	1,130		154
	10-15	13-19		880	
1975	below 15	below 15	1,400		182
	15-30	15-30		950	
1977	below 30	below 27	2,150		238
	30-50	27-45		700	
1979	below 30	below 22	2,400		357
	30-50	22-36		940	

Sources: 1968 to 1977, OECD, as for Table 4.3. 1979 reserves from OECD 1979, as reported in INFCE Working Group I, Summary Report of 5 November 1979. 1979 assessment of ten-year cumulative requirements from *The Balance of Supply and Demand 1978-1990* (The Uranium Institute, London, 1979). (Based on the Institute's more realistic low nuclear capacity assumption, and a 0.20 per cent enrichment tails assay.)

Third, it is significant that low-cost reserves increased steadily through the period up to 1975, during which they were defined by a relatively stable constant dollar price/cost limit. As could be expected, the quantity of reserves jumped sharply upwards in 1977, as a result of the increased price/cost limit in that year's assessment. More surprisingly, the tonnage went even higher in 1979, despite the return to a more restrictive price/cost limit in real terms, in defining low-cost reserves.

Fourth, each price/cost category of reserves consists of mineral deposits which can all be ranked according to the costs of their exploitation. Cost differences can be due to such factors as size of the deposit, amount of over-burden, ore grade, chemical complexity of the ore, geographical location and the availability of mining infrastructure. If total low-cost reserves are ample, exploitation will concentrate on the most economic deposits within the low-cost category.

Fifth, the industry has repeatedly stated that it needs an eight-year forward reserve for proper planning of its activities. To be on the safe side, we have compared the quantities of low-cost reserves with forecast ten-year forward requirements at different points in time. Our results in the last column of the table suggest that the amount of low-cost reserves has been more than adequate throughout the period and that it has grown faster than the forecast growth in requirements. The very high figures obtained for 1977 and 1979 are a clear indication that exploitation in the early 1980s can concentrate on the more economical deposits within the category, with extraction costs well below the nominal $30 limit.

Numerous claims have been made that the rate of growth at which new reserves are established in the US has been inadequate and falling, primarily as a result of a persistent fall in exploration productivity.[22] Most such claims, including the two quoted here, result primarily from a concentration of attention on the growth of $8 reserves in the US,[23] without taking the effects of inflation into account. It is indeed true that $8 reserves in the US, amounting to 150,000 short tons U_3O_8 in 1968, grew to 277,000 short tons in 1974, and then declined to 200,000 short tons in 1976, after which time they are no longer separately reported.[24] But since current dollars are used, a large proportion of the deposits formerly classified in the $8 category has had to be moved to higher-cost categories, as a result of inflation-induced increases in the costs of exploitation.

Table 4.5 provides data for US reserves organised in categories similar to those for the Western world in Table 4.4. It is seen that when constant dollar forward cost categories are used, the US reserves have risen very considerably from 1968 onwards, both in total magnitude and in relation to estimated ten-year forward requirements in the country. It may be noted that the constant dollar category used in Table 4.5 corresponds closely to the low-cost world reserves category in Table 4.4. In the US, like in the Western world, reserves have experienced an adequate growth and appear to be quite ample in relation to future uranium requirements.

Table 4.5: US Uranium Reserves

Year of Observation	Forward Cost Category, below		Amount, Short Tons U_3O_8	Reserves in Percent of Ten-year Forward Cumulative Requirements in the US
	Current $	Constant 1975/6 $		
1968	8	15.20	150,000	115
1971	10	15.80	300,000	145
1976	15	15.00	430,000	169
1979	30	21.60	690,000	255
1979	15	10.80	290,000	107

Note: Forecast US ten-year forward requirements in 1968, 1971 and 1976 have been approximated on the basis of linear growth from official forecasts providing annual requirements with five-year intervals. For 1976, the highest of several requirements forecasts has been used.

Sources: 1968, 1971 and 1976, US AEC and ERDA data as reported in *Uranium Price Formation*; 1979, *Statistical Data of the Uranium Industry* (US Dept. of Energy, Washington, DC, January 1979).

To add a further insight, we have included in the table the 1979 data for nominal $15 forward cost reserves. These reserves, although one-third cheaper to exploit in real terms, are twice as large as the nominal $8 reserves of 1968. Furthermore, they suffice by themselves to assure ten years' forward cumulative requirements in the US.

The claim of persistently falling productivity in exploration in the US is also mainly explained by the disregard of inflation. Exploration productivity can be defined as the amount of reserves in a constant dollar cost category, identified and established per million constant dollars in exploration expenditure. No reduction in this productivity can be discerned when reserves are categorised on the basis of constant dollars. A detailed recent study on the subject notes that 'there has been no significant decline in finding rate in the last 10 years, when expressed on a per-hole-basis, and using constant 1976 dollars'.[25] Another factor behind the claim of falling exploration productivity may be a misunderstanding of the relationship between the amount of drilling and reserve creation. This relationship is not a direct one. Drilling can be either for exploration to establish promising areas in general, or have the more detailed purpose of closely delineating formerly known ore bodies, about which further data are required. Ordinarily, only the latter will add to reserves, and as noted by Hogerton, 'it may be that the low yield figures of the past several years reflect the fact that some producers have lacked need or incentive to block out new reserves, and have directed their drilling programs towards other objectives.'[26] Finally, one must note the lag between exploration and reserve creation. Thus, it is probable that the very fast increase in exploration expenditure in the US since 1975 is resulting in substantial reserves which are not being recorded in official statistics until several years later.[27] For all the above reasons, the claim of falling exploration productivity in the US appears to be unproved.

Data on exploration from outside the US are insufficient to allow a similar evaluation of exploration productivity.

The conclusion emerging from our present discussion is that all through the period of low uranium prices, reserves did not constitute any constraint on current production nor on forward production planning, and that they were growing at quite adequate rates to meet forecast future demand. For the longer-term future it is worth noting that the uranium industry is young, and compared with most other minerals, exploration has barely started. The probability of further substantial economic discoveries is therefore reasonably assured.

An Analysis of Incentive Price Requirements

A new mining venture will be established only if the expected price covers not only operating costs, but also amortisation of used capital equipment and a return on invested capital. The most appropriate way to assess this return is through discounted cash-flow analysis. To explain how the discounted cash-flow rate of return is derived, we use a highly simplified chart of the anticipated income and expenditure flows of a hypothetical mine. In Chart 4.3, overall investment expenditures to bring the mine into production, occurring in years 0 to 6, amount to a total of $280 million. Operating costs in the subsequent period are estimated at $10 million per year until the cut-off at the time horizon in the 26th year. On the basis of expected prices, an annual income stream of $50 million, resulting from sales in years 6 to 26, can be determined. The discounted cash-flow rate of return is that rate which equalises the present value in year 0 of the overall expenditure and income streams. In our hypothetical example this rate of return is just below 10 per cent. The project will be developed only if the investors find this return on their capital to be satisfactory.

The discounted cash-flow rate of return is distinguished from the direct rate of return in that it takes account of the cost of capital in the pre-production period. The distinction is important, since this period is commonly quite extended in the mining industry. In our example, the direct rate of return would be above 14 per cent. Discounted cash-flow analysis is regularly used to determine whether planned investments are worth while. Except where specifically stated to the contrary, the rates of return presented below are all derived with the help of discounted cash-flow analysis.

Chart 4.3: Simplified Income and Expenditure Stream of a Hypothetical Uranium Mine

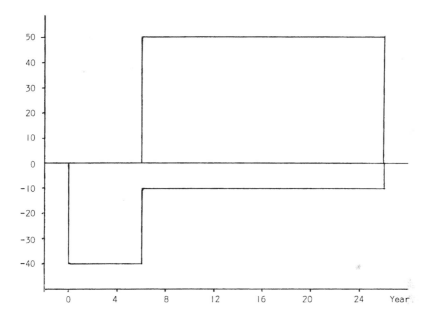

Chart 4.3 can also be used to clarify another issue. Once the investment has been undertaken, it becomes a sunk cost, and will cease to influence decisions whether to produce or not. In year 0, the investment decision was dependent on an anticipated annual income stream of $50 million. After year 6, when the investments have been completed, production will continue so long as gross annual revenue exceeds the $10 million needed to cover operating costs.[28]

We noted earlier in this chapter that the uranium price levels which prevailed in the 1968-73 period were sufficient to cover operating costs in most cases and thus to keep production going in existing installations. On the other hand, it was evident that as a rule prevailing prices were not adequate as inducements for new investments. The purpose of the next several paragraphs is to determine an incentive price level, thought necessary in the early 1970s to assure a growth in capacity and supplies commensurate with forecast requirements.

Since the non-US evidence for our task is not as plentiful as required for a detailed analysis, we have primarily used data of US origin. An interesting point of departure is the finding that the 17 US uranium mining/milling producers operating in 1974 had reported an average direct rate of return on total assets of 5.6 per cent before tax for the entire period of 1965-74,[29] during which the 1975/6 dollar uranium prices averaged $12.25.[30] Since we know neither the asset figure nor how it was calculated, we cannot draw any far-reaching conclusions from the above profit figure. However, if we assume that 90 per cent of the revenue received over these years was needed to cover costs and the remaining 10 per cent of revenue provided for the stated return, it is possible to infer that the direct rate of return on total assets (before tax) in these firms would have been 32 per cent at a price level of $18 (1975/5 dollars). This figure appears high enough as an inducement for investments in new ventures.

A far more reliable approach to the determination of incentive price levels can be based on the data contained in a paper written in 1972 by John Klemenic[31] of the US AEC. In this paper, Klemenic presents examples of total costs in future uranium-mining operations. In order to bracket the range of possible industry conditions, the author undertook sensitivity analyses on his base case, by assuming varying rates of operations, ore grades, exploration costs, etc. Klemenic's data were based on a wide array of actual company costs, and have been repeatedly used in assessments of incentive prices.

Analysing the Klemenic data with the help of a computer model, Al Petrick concludes[32] that prices of between $9.05 and $13.17 per lb U_3O_8 (of 1972 dollars) would be required to assure a 12 per cent after-tax discounted cash-flow rate of return to the various hypothetical mining ventures considered by Klemenic. Hogerton, too, has used the Klemenic data, and concludes that in an 'average' venture, an incentive price of $10.50-10.70 per lb (1972 dollars) is needed to assure a 15 per cent discounted cash-flow rate of return.[33] In another report, Hogerton[34] refers to a paper by D.A. McGee, the chief executive of a major US uranium-producing firm, analysing the required price to provide a 10 per cent after-tax discounted cash-flow rate of return to a low-grade underground uranium deposit. McGee concludes that the required price (in 1971 dollars) would have to be $10.40. D.O. Cooper, in a frequently quoted paper,[35] concludes that a viable price in 1972 dollars, assuring an 11 per cent return on investments in an average venture, would be between $10 and $11.

Petrick[36] also scrutinises consultants' reports evaluating specific mining ventures in the US in the early 1970s. He concludes that the incentive prices needed to get the prospects developed fall within the range obtained on the basis of Klemenic's data. Further scrutinies of several mining prospects in Canada and Australia indicate required incentive prices at the lower end of the price range derived for the US, and in some cases considerably below that range.

The above views of the early 1970s can be summarised as follows. Expressed in constant dollars, the Petrick/Klemenic upper-range figure (1972 $13.17) is the highest of the incentive prices quoted, and we feel it covers the upper limit of contemporary opinions about the prices required to bring about an expansion in the uranium industry commensurate with forecast requirements. Using our price deflator, this figure works out at $19 in 1975/6 dollars. This can be compared with the actual average price in 1973, expressed in 1975/6 dollars, of $8.10, and with the average prices in 1975 and 1976 of $24 and $39 respectively (NUEXCO spot).

The need to set an incentive price as high as Petrick/Klemenic's $19 (1975/6 dollars) in the mid-1970s would depend on the need to develop the marginal US deposits to which this price applies. Incentive prices could be set lower if, for instance, a sufficient quantity of new and more economical deposits were identified within and outside the US, or if nuclear generating capacity expansion was such as to permit the satisfaction of overall uranium requirements without the need to bring the above marginal deposits into production.

An approach to required uranium prices very different from ours has been proposed in several studies of the uranium industry.[37] The argument relates to the US conditions in particular, and assumes that the prices paid to US producers by the US AEC in the 1950s fairly reflect the prices required to induce the industry to expand. Taking it that costs have increased in proportion to the fall in the ore grade over the past decades, and adjusting prices/costs for ongoing inflation, the authors claim that required price levels in current dollars ought to be close to $40 in the mid-1970s. Hence, the studies suggest that only in 1976 did prices reach levels assuring adequate growth of the uranium industry.

In our opinion, the above analysis rests on two faulty foundations, and must be refuted. *First*, it fails to consider the improvement in productivity experienced by the uranium industry over the third quarter of the present century. Our discussions earlier in this section noted that improving productivity has succeeded in countering the upward cost push from gradually reduced ore grades in several established minerals industries over long periods of time. It is reasonable to argue that productivity change must have had a stronger impact on uranium exploitation than on the minerals sector as a whole. On top of the general improvements accruing to the entire mining sector, uranium productivity should have benefited strongly because the industry started the period as an infant industry, with small production units, and without experience. As it gradually matured, substantial productivity gains have undoubtedly been derived from increasing scale of operations and from learning new techniques in producing this mineral.

The above argument is impossible to document without a monumental empirical investigation. Scattered inferences from the uranium literature can be quoted to substantiate the claim that productivity has improved. The UK paper to the 1970 Foratom Congress,[38] for instance, states that costs in uranium-mining in North America had fallen by something like 40 per cent between 1956 and 1970, when measured in constant dollars, despite the simultaneous fall in the ore grades. It is significant that one of the paper's authors represents a leading uranium mining company. In the US,[39] in the earlier periods, hand-mining methods had to be employed in the small units then in production. Only gradually was it possible to introduce cost-reducing large-scale mechanical equipment. The size of mines has increased throughout the period. From 1968 till 1975, for instance, the average mine size grew by over 60 per cent in terms of U_3O_8 production, with a significant impact on cost levels. Labour productivity in terms of U_3O_8 in US uranium mills appears to have doubled between 1959 and 1970, while overall milling costs experienced a fall by more than half from 1960 till 1972, when measured in real terms. This discussion and evidence suggests that it is far more reasonable to assume that improving productivity has at least neutralised the upward cost push due to falling grades in uranium ore, than to assume away productivity changes altogether.

The *second* fallacy of the argument adopted in the quoted studies is that the prices paid by the AEC to US producers in the 1950s represent reasonable incentive levels needed to ensure a fast and steady growth of the uranium industry. This is simply not true. As detailed in Chapter 3, the AEC prices constituted the successful backbone of a veritable crash programme. The price incentives were so great that they induced US producers to increase output from 810 short tons in 1950 to 19,940 short tons in 1960, or by a factor of 23.[40]

A study of Table 3.1 reveals that the constant 1975/6 dollar prices paid by the AEC to US producers rose from $30 in 1950 to $35 in 1953, and then fell gradually to $21 in 1960. The average price paid over the whole decade was $29. It is not entirely unreasonable to conjecture that world producers, unhampered by political constraints, could repeat the achievement of US producers, if they were reasonably assured of similar real prices over a whole decade. A 23-fold production increase, even if starting from the restricted 1975 figure of 25,000 short tons, would then bring 1985 production to 575,000 short tons, wildly above any forecast requirements.

The purpose of the above exercise is not to show what could happen; it is merely to point to the special nature of the prices paid by the US authorities in the 1950s. It is unreasonable to compare those prices (in real terms) with the incentive prices needed by the world uranium industry in the 1970s and 1980s.

Price Forecasts for the 1970s and 1980s

We now turn to the forecasts and predictions about future uranium prices made at various times between the late 1960s and 1974, i.e. until the time when the prices started their sharp rise. One should be careful to distinguish between opinions on the level of required incentive prices on the one hand, and forecasts of market price developments on the other. For a variety of reasons, actual or forecast market prices can deviate strongly from the levels needed to induce producers to expand supply in line with developing needs.

Price forecasts are always heavily influenced by current prices, and by the price history of the recent past. Thus, just as many of the forecasts made in the late 1970s tended to project price levels of $40 or more into the future, there was a widespread view in the late 1960s and early 1970s that uranium prices in the next five to ten years would rise only marginally, or not at all. This view, naturally, derived support from the excess capacity, surplus production and rising stocks, characterising the period. Lorie and Gody,[41] in their paper, have a number of quotations reflecting forecasts of this kind.

The alternative major view of future prices, while also noting the current surplus situation, was more impressed with the expectations of very fast-growing uranium needs. The resulting price forecasts foresaw a turnaround in the market about the time when converging production capacity and requirements forced the industry into undertaking new investments on a large scale. Some of the proponents of this view have been referred to in an earlier section of this chapter. However, neither of the views led to predictions of uranium prices anywhere near $40 (1975/6 dollars) during the 1970s and 1980s.

In providing evidence on opinions about future uranium prices expressed in the period under scrutiny, we propose to be selective, first, by excluding that primarily US evidence already presented in the report of Lorie and Gody, to which the interested reader is referred, and, second, by concentrating on forecasts which appear to be based on a thorough and independent analysis, not merely repeating the opinions of others.

First, and probably most important in our list of price forecasts, is the weight of opinions about future prices as expressed in long-term contracts entered into in the early 1970s. Table 4.6 reflects the prices of contracts signed in 1971-3, recorded by Euratom (with buyers in the EEC) and by NUEXCO (predominantly in the US), for deliveries between 1977 and 1980. It is clear from the table that even as late as 1973, the contracting producers did not expect 1980 prices in current dollars to be much above $10. If higher prices had been expected, a rational policy for the producers would have been to abstain from entering into the future sales commitments until the time when the expected higher price could be obtained.[42]

Table 4.6: Uranium Prices for Future Delivery ($/lb U_3O_8)

Time of Contract Signature	Price for Indicated Year of Delivery				Source
	1977	1978	1979	1980	
1971			7.50		Euratom 1971
End 1971		6.00-6.60			Euratom 1972
Mid-1972	7.15-7.40	7.50-7.80			Euratom 1972
1973		7.60-10.00	7.90-10.40	8.10-10.40	Euratom 1973[a]
Mid-1972	8.30				NUEXCO
Mid-1973	8.50	9.00	9.50	10.00	NUEXCO

Note: [a]Prices per lb U_3O_8 including cost of conversion into UF_6.

Sources: Euratom Versorgungsagentur, *Annual Reports*, as indicated, and regular monthly NUEXCO reports.

Our remaining sources of evidence are organised chronologically. To facilitate comparison, we provide the forecast price levels, wherever possible, in terms of constant 1975/6 dollars.

Mårtensson,[43] reporting the consensus view of a Nuclex meeting during *autumn 1966*, noted that prices of $5-6 were too low, but that it should be possible to maintain future prices below the $10 level. It is not clear whether this related to current or constant dollars; neither is any time perspective given for the forecast.

Sigvard Eklund[44] (Director General of the International Atomic Energy Agency) in *1968* claimed that uranium prices would increase from the current level of about $6 to between $13 and $15 (1975/6 dollars) during the 1970s, if no substantial new uranium deposits were discovered.

In *March and April of 1969*, the UK Ministers of Technology and Power stated[45] that according to expert opinion in the UK no significant increases in the price of uranium were foreseen over the coming five years, and that the high levels of ongoing exploration were seen to assure sufficient uranium supplies at reasonable prices beyond that time period.

Writing in *1970*, Houdaille[46] (Director General of Uranex, the French uranium marketing agency), concluded that excess capacity would characterise the uranium market at least until 1976-8, with depressed prices up to 1975. Between 1975 and 1977, he expected a price level of about $11 (1975/6 dollars). Beyond 1977, new capacity would be needed, and one could expect prices above $13 (1975/6 dollars), with variations of a dollar or two around their average level.

In *1970*, the French Commissariat à l'Energie Atomique[47] forecast prices of $13-14 (1975/6 dollars) in the mid-1970s, and no higher than $13-17 (1975/6 dollars) in the 1980s. It added that competent French authorities believed in a calm and stable development of the uranium market.

The UK contribution to the *1970* Foratom Congress in Stockholm[48] noted a widespread agreement among uranium suppliers and buyers that prices would not rise above the level needed to cover production costs and reasonable profits. In 1970, this level was stated to be about $8. While suppliers felt that the level might rise in real terms 1-3 per cent per year, buyers doubted whether it need increase at all. With the high extreme of a 3 per cent increase in real terms, the 1970 $8 price works out at $18 (1975/6 dollars) by 1980.

The Swedish contribution to the same congress (*1970*)[49] contained no price forecast, but claimed that 330,000 short tons of U_3O_8 could be exploited on a large scale from a meagre Swedish deposit at prices of $17-25 (1975/6 dollars). Given the size and marginal nature of the deposit, the prices required for its exploitation could safely be regarded as an upper price limit unlikely to be surpassed in the foreseeable future.

The Swedish State Power Board, writing in October *1971*, felt that the current uranium prices were too low, but believed that the future price would not exceed $16 (1975/6 dollars) before the end of the 1980s.[50]

Johan Brinck[51] of the Directorate General for energy in the European Economic Community, writing in *1971*, contended that existing reserves and production capacity were sufficient at least up to 1977/8, but that prices might rise slowly towards $12-13 (1975/6 dollars) in the late 1970s.

In an analysis of future uranium-mining prospects undertaken by Charter Consolidated in *1971*, the long-run uranium price (late 1970s and early 1980s) was forecast at about $12 (1975/6 dollars).[52]

In *September 1972*, J.F. Hogerton[53] forecast that the then prevailing current price of around $6 would soon experience a significant, or even substantial, increase, to reach somewhere between $6.50 and $8. From then on, Hogerton predicted a continued, though unspecified, price rise.

Hogerton's forecast of *November 1973* was more precise.[54] Up to 1985, uranium prices were expected not to rise much above $14 (1975/6 dollars).

In a large study completed in *December 1973*, General Electric[55] forecast a fast uranium price increase, to peak in 1977 at $16 (1975/6 dollars). From then on, prices were forecast to fall to $15 (1975/6 dollars) in 1980, and to $14 (1975/6 dollars) in 1985.

Analysing the prospects of a uranium-mining company, D.S. Robertson & Associates Ltd[56] in *July 1974* forecast uranium prices to rise rapidly, to reach $15 (1975/6 dollars) some time between 1980 and 1982.

Interestingly, the last two forecasts, made after the sharp oil price increases which occurred in 1973, do not project much higher prices for the late 1970s than the forecasts made before the oil crisis. Even in late 1975, after two years with high oil price levels, Hogerton forecasts[57] uranium prices for 1985 at between $16 and $20.50 (1975/6 dollars), i.e. no more than 15-45 per cent above his forecast of November 1973 and substantially below current NUEXCO spot quotations. This suggests that during 1974 and 1975 influential uranium market observers believed either than the oil prices would fall in the near future, or that they did not see any strong connection between uranium and oil prices.

We summarise the above evidence by noting that in the 1971-3 period, the uranium market itself was by far the most pessimistic (or perhaps the least far-sighted) about future prices. Prices for the end of the decade were set as low as $8-10 in *current* dollar values. This was substantially below the forecasts made between 1971 and 1973 by experienced observers and institutions. The latter foresaw a rise in prices to perhaps $12-14 in fixed 1975/6 dollars for the late 1970s and early 1980s.

Inflation expectations were not very high in the early 1970s. It is clear that the prices for the end of the decade stipulated in the contracts of 1971-3 were seen to be not much higher in real terms than current prices at the time of contract signature. The willingness of market agents to enter into such arrangements suggests any of three possibilities:

(1) that the buyers and sellers of uranium were ignorant about the forecast long-term growth of uranium demand, and therefore did not consider the implications on price of required investments in new capacity;

(2) that the buyers and sellers were aware of the forecasts but did not believe they would materialise;

(3) that neither buyers nor sellers took their long-term contractual obligations very seriously.

Though in retrospect contract conditions proved quite amenable to manipulation, we find it hard to believe that the third possibility could have been an important factor in the determination of price levels, and conclude that the stagnant future prices in the long-term contracts of that time resulted from a combination of the first and second statements.

The comments attached to each of the forecasts presented above suggest that the forecasters, as distinct from the market agents, took serious notice of the implication of the fast growth in uranium requirements which was expected to bring demand up to the level of existing production capacity by the mid- or late 1970s. The very substantial price rises contained in their predictions represented their opinions on the incentive price levels seen to be necessary to induce the additional investments in new capacity to satisfy the growing demand.

There is no contradiction between the forecasts of 1971-3, contending that uranium prices would reach $12-14 (1975/6 dollars) by the end of the decade and the contemporaneous incentive price exercises discussed earlier in the chapter. The $19 (1975/6 dollars) incentive price quoted by Petrick refers to high-cost marginal mining ventures which the price forecasters did not take into account, because they believed demand growth could be amply satisfied from more economical, lower-cost deposits.

Nowhere in our survey of literature and analysis of what was happening in the uranium market up to 1973/4 have we been able to find any indication of impending conditions that might lead anywhere near to the price explosion which actually occurred. It is therefore not surprising that we failed to find any foresights of the price developments that were coming. On this basis we feel it reasonable to conclude that the uranium market developments between 1973 and 1976 were unexpected and that, in all likelihood, the factors which brought about the drastic change must be sought elsewhere than among 'normal' forces operating in that market.

The Paradox of Depressive Market Factors and Explosive Prices after 1973

So far in this chapter, the emphasis has been on developments in the uranium market until about 1974, when the price explosion began. On several occasions when presenting data series, we extended the coverage and discussion to more recent years. It appears that the factors which would ordinarily be expected to determine the demand, supply and price of uranium suggest continued depressed conditions in this market long after the prices had in fact exploded. In this concluding section we survey in a more systematic manner the development of these factors through the period of rising and high prices.

We begin by looking at the forecasts of installed nuclear power capacity, the key determinant of uranium demand. Chart 4.4 reproduces in somewhat greater detail the contents of Table 4.2. The 1975 and 1977 forecasts are given as bands, representing the high and low estimates respectively, made in each year.

Forecasts of nuclear power growth reached an all-time maximum in 1970, and have been persistently declining ever since. The chart presents four consecutive OECD forecasts, the latest dated December 1977. Since then a further decline has taken place. Three long-run forecasts made in 1979 assess the future nuclear capacity build-up in bands around or below the lower OECD forecast of 1977. Thus, INFCE projects the 1990 capacity to be between 370 and 460 GWe, and the 2000 capacity at 850-1200 GWe.[58] The Uranium Institute projects the installed capacity in 1990 at between 410 and 530 GWe, with the lower

Chart 4.4: Forecasts of Nuclear Power Capacity Installations, Western World

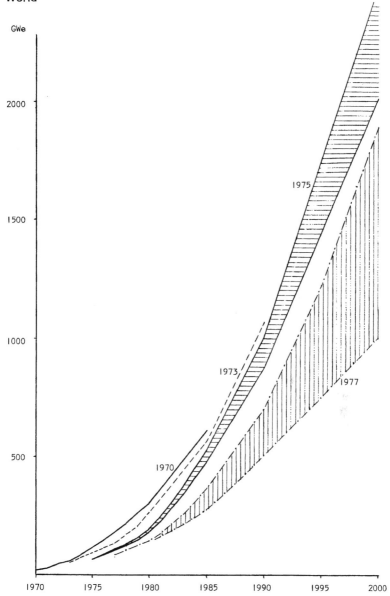

figure considered to be more realistic,[59] while the Australian Atomic Energy Commission forecasts a lower band of 350-440 GWe for the same year.[60]

Three factors provide the major explanation to the reduced nuclear power forecast figures. First, nuclear reactor construction has frequently been delayed, primarily by the protracted regulatory process in many countries. The regulatory requirements have become an increasingly serious obstacle to speedy investment execution over the years. In the US, for instance, the average construction period increased from 43 months for reactors which became operative in 1970 to 96 months for reactors completed in 1979.[61] Second, the long-term electricity demand expansion has been scaled down as a result of higher prices and the more gloomy economic growth prospects in OECD countries in the post-1973 era. Finally, and probably most important, there has been a mounting opposition in many countries to nuclear power, for environmental and security reasons, culminating with the Harrisburg accident of 1979.

Contrary to widespread belief, the oil crisis of 1973/4 did not lead to expanded efforts world-wide to install nuclear power. As appears from the chart, even the higher estimates of 1975 are persistently below the 1973 forecast figures. France was one of the few countries where plans for a more speedy nuclear programme were adopted in consequence of the higher oil price and insecurity of oil supply after 1973. In all other major countries, the nuclear plans were reduced, despite the developments in the oil market.

The reductions in planned nuclear capacity have been very substantial. Between 1973 and 1977, for instance, the projected average capacity figures for the 1978-82 period declined by 46 per cent.[62] This decline obviously had a strong impact on the forecast needs for uranium. Starting out from the 1973 OECD projection of uranium requirements for the 1978-82 period, 380,000 short tons, and assuming that this quantity declined in line with the falling nuclear capacity forecast, we can deduce that, everything else alike, the requirements would have been no more than 206,000 short tons in the same five years on the basis of the 1977 projection for nuclear expansion. This corresponds to a reduction in average annual requirements of some 35,000 short tons over the five-year period considered.

Uranium prices were still low at the time when the 1973 forecast was presented, and one would hardly have expected them to rise on the basis of so substantial reductions in potential needs in the subsequent years.

The sharp fall in reactor ordering in 1975 should have been of even greater and more immediate concern for the uranium market than the long-term capacity generation forecasts. Western world annual reactor orders are listed in Table 4.7.[63] The figures clearly constitute a strongly depressive factor in the uranium market from 1975 onwards.

Table 4.7: Western World Annual Reactor Orders

Year	No.	MWe
1970	27	22,400
1971	37	35,400
1972	61	59,400
1973	60	65,400
1974	67	68,700
1975	16	17,800
1976	33	33,700
1977	12	n.a.

A second important factor is the perseverance, at least until 1977, of uranium production capabilities well above current uranium needs. The gap between capability and requirements, as given by the OECD, and detailed in Table 4.3, has been so large that it cannot possibly all be explained away by the uncertainties which always surround capacity data. On this evidence it would be hard to argue that capacity constraints contributed to the factual price performance in the uranium market after 1973. On the contrary, the continued excess capacity should have contributed to a maintenance of depressed prices, at least until 1977.

Third, it is clear from a comparison of production and requirements figures as detailed in Table 4.3 that in every year up to 1977, the former was above the latter. Data for 1978 indicate that this was true also for that year.[64] This logically leads to the conclusion that there must have been a continuous build-up of inventories. Figures on uranium stocks are not regularly published in most countries. However, Table 4.8 summarises data for the US between 1973 and 1977. Requirements in the US were estimated at about 13,000 short tons in 1977.[65] Hence, users' stocks alone were sufficient to cover more than two years' current nuclear reactor needs. And evidence of contractual commitments to buy suggested that these stocks would continue to increase up to the mid-1980s.[66] In Germany, government stocks in 1977 were reported at

3,200 short tons, up from 1,800 short tons in 1975, while users' stocks, reported to have grown fast to reach 2,000 short tons in 1977, were adequate for more than one year's requirements.[67] The Uranium Institute's assessment of 1978 uranium stocks in the Western world results in a figure of 130,000 short tons $U_3 O_8$, which the Institute states is adequate to cover four years forward requirements.[68] The Uranium Institute figure excludes:

inventories held in intermediate production stages (mainly in enrichment plants);
stockpiles held for military or other non-commercial purposes;
depleted uranium from enrichment plants;
uranium awaiting delivery or in transit from producers.

Though available information is far from complete, the scattered data presented here indicate that inventories are large and growing. One would not ordinarily expect a price explosion in a commodity market where production has, for a long time, regularly exceeded requirements and where, consequently, the stock levels experience a continuous rise, all while projections of future demand are declining.

Table 4.8: Uranium Stocks in the US (short tons $U_3 O_8$ equivalents)

	Jan. 1973	Jan. 1975	Jan. 1977
Producers	3,700	4,300	2,400
Users	14,400	19,500	29,300
Government	50,000	71,500	78,800
Total	68,100	95,300	110,500

Source: OECD, *Uranium Resources, Production and Demand* (August 1973); ibid. (December 1975); ibid. (December 1977).

Finally, the data contained in Tables 4.4 and 4.5 indicate unambiguously that low-cost uranium reserves grew faster than cumulative future requirements of uranium between 1973 and 1975, and strongly suggest that the same pattern continued in more recent years. Any constraint on future supplies exerted by limited reserves, which may have been felt in 1973, must consequently have weakened in the following years.

The four fundamental factors discussed in this section, i.e. the emerging prospects for demand growth, the persistence in excess capacity, in excess production and in the build-up of inventories, and the fast expansion of economically favourable reserves after 1973, all suggest that the market should have remained in a depressed state. One must therefore look beyond the normal factors which determine demand, supply and price in commodity markets in order to explain satisfactorily why in fact the prices rose so sharply. We venture into this task in the next few chapters.

Notes

1. *Mineral Facts and Problems*, 1975 edition, US Bureau of Mines Bulletin 667 (Washington, DC, 1976).
2. J.P. Langlois, 'The Uranium Market and its Characteristics', *Uranium Supply and Demand* (The Uranium Institute, London, 1978).
3. See *Uranium Supply and Demand* (1978), pp. 21 and 49.
4. Langlois, 'The Uranium Market and its Characteristics'.
5. G. White of Energy Services Company, a division of NUEXCO, makes this claim in his paper 'Ten Year Forecast of Price Trends in the Domestic Uranium Industry' (July 1972).
6. *Uranium Price Formation*, EPRI (October 1977), prepared by Charles Rivers, Cambridge, Mass., p. 5. 26.
7. OECD, *Uranium Resources, Production and Demand* (OECD, Paris, August 1973). This publication is one in a series issued by OECD, to which frequent reference will be made. The full titles in the series are as follows:

OECD, *Uranium Resources*, (OECD, Paris, December 1967);
OECD, *Uranium Production and Short-term Demand* (OECD, Paris, January 1969);
OECD, *Uranium Resources, Production and Demand* (September 1970);
OECD, *Uranium Resources, Production and Demand* (August 1973);
OECD, *Uranium Resources, Production and Demand* (December 1975);
OECD, *Uranium Resources, Production and Demand* (December 1977).

In the following we refer to these publications as OECD, stating the appropriate date.
8. Ibid.
9. Production in the Australian Ranger and Nabarlek deposits was estimated in 1971 to cost $5.08 and $3.38 per lb U_3O_8 respectively, including a 15 per cent allowance for amortisation and depreciation. Average price in the same year was $6.25. See W.F. Atkins of D.S. Robertson & Associates Ltd, 'Production Costs for Three Uranium Properties, Northern Territory of Australia' (December 1971).
10. For a discussion of the delays in anticipated reprocessing and plutonium use, see Chapter 5.
11. OECD (September 1970).
12. OECD (August 1973).
13. For views of this kind see, for instance, D.D. Bell, 'Nuclear Fuel Resources and Price Trends', *The Canadian Mining and Metallurgical Bulletin* (April 1967); M. Mårtensson, 'Uran för Kraftalstring', *Teknisk Tidskrift* (1967), H 40; S. Eklund, 'Kärnenergi', *Ymer* (1968), pp. 246-58; Merrill Lynch *et al*. Inc., 'The Canadian Uranium Industry, an Institutional Study' (July 1972); J.F. Hogerton

et al., the S.M. Stoller Corporation, 'The Uranium Supply Outlook' (September 1972); D. Robertson & Associates Ltd, 'Uranium Supply and Demand', paper prepared for the Advisory Committee on Energy (Province of Ontario, Canada, 1972); O.J.C. Runnals, Dept. of Energy, Mines and Resources, Canada, writing in *Geos* (Fall 1972); General Electric, 'Uranium 1973-1985, Materials Resources Planning Report' (December 1973).

14. See, for instance, J. Kostuik, President Denison Mines Ltd, 'Key Issues Affecting the Future Development of the Uranium Industry', *Uranium Supply and Demand* (The Uranium Institute, London, 1976).

15. J. Klemenic, US ERDA, 'Uranium Supply and Associated Economics: a Fifteen Year Outlook' (October 1975), p. 5.

16. While Western world capacity is assessed by OECD in 1975 at 33,000 short tons, a simultaneous US ERDA estimate gives a figure of 37,000 short tons. See *Statistical Data of the Uranium Industry* (ERDA, January 1975).

17. R. Prain, Chairman Roan Selection Trust, 'The Future Availability of Copper Supplies', speech at the Institute of Metals, Amsterdam, 1970.

18. Metallgesellschaft, *Metal Statistics*, 1960 and 1977.

19. In the US in the early 1970s, about 3 lb of U_3O_8 were extracted from each ton of ore. World-wide, the figure was much higher.

20. For a full discussion of the reserves concept, the development of reserves over time and the meaning and significance of depletion in some common minerals, see M. Radetzki, 'Metal Mineral Resource Exhaustion and the Threat to Material Progress, the Case of Copper', *World Development* (February 1975), and M. Radetzki, 'Will the Long-run Global Supply of Industrial Minerals be Adequate? A Case Study of Iron, Aluminium and Copper' in Christopher Bliss and Mogens Boserup (eds.), *Economic Growth and Resources*, Vol. 3, *Natural Resources* (Macmillan for the International Economic Association, London, 1980).

21. For evidence see the two references in the previous footnote. Also see O. Herfindahl, *Copper Costs and Prices 1870-1957* (Johns Hopkins, Baltimore, 1959).

22. See, for instance, M.A. Lieberman, 'US Uranium Resources, an Analysis of Historical Data', *Science* (30 April 1976), or D. Robertson & Associates Ltd, 'Uranium Price Movement and the Reasons Therefor' (31 August 1977).

23. Reserves in the US are categorised on the basis of 'forward costs', i.e. all the remaining costs which have to be incurred to convert reserves into final product. Interest on capital and profit are not included in forward cost, however.

24. US ERDA data as reported in *Uranium Price Formation*, p. 5.4.

25. Pickard Love & Garrick Inc., 'Natural Uranium, Demand, Supply and Price' (February 1977).

26. J.F. Hogerton *et al.*, of S.M. Stoller Corporation, 'Report on Uranium Supply' (5 December 1975).

27. J.A. Patterson, 'US Uranium Supply and Demand Overview', paper presented at a meeting of the American Nuclear Society, 24 January 1977.

28. This assumes that the producer is a price taker and operates at full capacity.

29. National Economic Research Associates Inc., 'Competition in Uranium and Coal Markets, with Special Reference to Oil and Gas Companies' (New York, June 1979).

30. *Uranium Price Formation*, p. 9.2. The nominal prices were deflated by our price index, as presented in Chapter 2.

31. J. Klemenic, 'Examples of Overall Economics in a Future Cycle of Uranium Concentrate Production for Assumed Open Pit and Underground Mining Operations', AEC Nuclear Industry Seminar, Grand Junction, Colorado, 1972.

32. Al Petrick, 'An Economic Analysis of the Relationship Between the Cost and Price of Uranium' (Petrick Associates, 1979).

33. Hogerton *et al.*, the S.M. Stoller Corporation, 'The Uranium Supply Outlook' (November 1973), p. 36.

34. Hogerton *et al.*, 'The Uranium Supply Outlook' (September 1972), p. 44.

35. D.O. Cooper of Getty Oil, 'Viable Uranium Exploration and Production Economics', Atomic Industrial Forum Seminar on Uranium, March 1973.

36. Al Petrick, 'An Economic Analysis of Uranium Prices' (draft, 1979).

37. Variations of this approach can be found in Robertson, 'Uranium Price Movement and the Reason Therefor', in a note by D.M. Johnson of United Nuclear Corporation, entitled 'Observations on Uranium Prices'(24 November 1975), and in M. Mårtensson, AB Atomenergi (Sweden), 'Ekonomisk Analys av Uranmarknadens Historiska Utveckling' (July 1977).

38. J.S. Clarke, P.J. Searby and L.C. Mazel, 'UK Uranium Demand and Procurement Strategy', paper presented at the 1970 Foratom Congress, Stockholm.

39. The rest of this paragraph is based on *Uranium Price Formation*, Chapters 6 and 7. The study quoted notes a sudden reduction in labour productivity in uranium-mining in the US in the mid-1970s, ascribes it primarily to the fast influx of inexperienced workers into the industry at this time, and considers it to be a passing phenomenon.

40. See Chart 3.1.

41. J.H. Lorie and C.S. Gody, 'Economic Analysis of Uranium Prices', a report prepared for Westinghouse Electric Corporation (9 July 1975).

42. We are well aware of the fact that producers sometimes have to sell at distress prices to maintain required cash flows. This does not conflict with our argument, however. We do not claim that producers should have restricted their sales from current production, or refrained from entering into delivery contracts for the nearby future. We merely suggest that if higher prices had been expected in the late 1970s, it would have been rational for the producers to defer committing their supplies for such future years until prices had adjusted to their expectations.

43. Mårtensson, 'Uran för Kraftalstring', H 40. Mårtensson also notes a paper by G.R. Ball, 'World Uranium Supply and Demand Factors as Applied to an Econometric Model, with Special Reference to Stockpiling', later published in *Neue Technik* (February 1967), which forecasts a price rise from $10 in 1974 to $20 in 1977 (unclear whether current or constant dollars). Mårtensson agrees with the general opinion of nuclear experts that Ball fails in properly presenting his assumptions and that his price predictions appear exaggerated. In our view, the results obtained by Ball cannot be judged either way, since the econometric model from which they are said to be derived is never presented.

44. Sigvard Eklund, 'Kärnenergi', *Ymer* (1968), pp. 246-58.

45. *Atom* (May 1969 and June 1969).

46. M. Houdaille, 'Le Marché de l'uranium', *Bull. Inform. Ass. Tech. Energ. Nucl.*, no. 84 (July-August 1970).

47. *L'industrie minière de l'uranium* (Republique Française, Commissariat à l'energie atomique, 1970), pp. 107-8.

48. Clarke, Searby and Mazel, 'UK Uranium Demand and Procurement Strategy'.

49. R. Gehlin of AB Atomenergi, Stockholm, 'Uranium Demand and Supply', paper presented at the 1970 Foratom Congress, Stockholm.

50. Swedish State Power Board Memo. ER – J11/KW – 6320, dated 7 October 1971.

51. J. Brink, 'Uranium Prospects and Problems', *Eurospectra*, no. 1 (1971).

52. *Metals Analysis and Outlook*, no. 2 (August 1976) (Charter Consolidated Ltd, London).

53. Hogerton *et al.*, 'The Uranium Supply Outlook' (September 1972).

54. Hogerton *et al.*, 'The Uranium Supply Outlook' (November 1973).

55. General Electric, 'The Uranium Market 1973-1985' (December 1973).

56. D.S. Robertson & Associates Ltd 'Evaluation of Allied Nuclear Corporation' (July 1974).

57. Hogerton *et al.*, 'Report on Uranium Supply' (5 December 1975).

58. See report on the work of the International Nuclear Fuel Cycle Evaluation in *Nuclear Fuel* (20 August 1979).

59. *The Balance of Supply and Demand 1978-1990* (The Uranium Institute, London, 1979).

60. Australian Atomic Energy Commission, *Annual Report* for year ended 30 June 1979.

61. K.R. Shaw, 'Capital Cost Escalation and the Choice of Power Stations', *Energy Policy* (December 1979).

62. OECD (August 1973 and December 1977).

63. T.L. Neff and H.D. Jacoby, *Nuclear Fuel Assurance, Origins, Trends and Policy Issues*, MIT Energy Laboratory Report No. MIT-EL 79-003 (MIT Press, Cambridge, Mass., February 1979).

64. *The Balance of Supply and Demand 1978-1990.*

65. J.F. Hogerton, 'US Uranium Requirements' in *Uranium Supply and Demand* (The Uranium Institute, London, 1977), p. 46.

66. J.F. Hogerton, 'Disturbing Conflict of Views on Uranium Supplies' *Nuclear Engineering International* (November 1977).

67. *Atomwirtschaft* (December 1977); OECD (December 1975); and K.P. Messer, 'Uranium Demand as Judged by Electric Utilities' in *Uranium Supply and Demand* (1977), p. 56.

68. *The Balance of Supply and Demand 1978-1990.*

5 THE POLITICS OF REPROCESSING AND ENRICHMENT

Introduction

The major purpose of this chapter is to analyse how a set of policies in the field of reprocessing and enrichment, proclaimed from 1973 onwards, added to the demand for natural uranium. The public bodies which formulated the policies appear to have had little concern about the ensuing economic implications. Hence, it seems reasonable to label the decisions political, even though there may have been implicit economic or commercial considerations in the background.

Three issues are brought into focus. The first one concerns the delays in establishing and operating spent fuel reprocessing capacity. The second deals with the enrichment tails assays. The third one pertains to the contractual arrangements in uranium enrichment.

The military heritage gave the US a technological lead in all phases of uranium processing. Only in the late 1970s has the complete dominance of US public institutions in this field started to break up. Until then, the politics as well as economics of reprocessing and enrichment of uranium lay firmly in US hands. For this reason, the chapter deals primarily with the policies of the US government in the fields of reprocessing and enrichment.

Delays in Reprocessing

Reprocessing of spent fuel elements to extract the remaining uranium and the plutonium formed through the fission process can provide for almost one-third of the fuel needs of light water reactors. Though the technology has been developed, and reprocessing is claimed to be economical, very little spent fuel has been recycled so far. This is mainly due to the fact that the US authorities have repeatedly delayed permission to construct and operate uranium fuel reprocessing plants in that country. The delays have been partly due to environmental and regulatory problems but also to technical difficulties in running spent fuel recovery facilities. To illustrate, in 1968, reprocessing was expected to start in the mid-1970s. In 1974, after some delays had occurred, it was still believed that full recycling would be possible by 1980.[1] By 1975 it became clear that reprocessing would take much longer to introduce, and in April 1977, finally, the US President announced an indefinite postponement of reprocessing in the US, and an embargo on exports of the relevant technology,[2] to delay the establishment or expansion of recycling in other countries. A major motivation for this policy move appears to have been the danger perceived by the President and his associates of the plutonium economy, in which large amounts of this lethal material are manufactured, transported and used in the course of ordinary commercial transactions.

Uranium fuel recycling has been slow in developing in other countries too. In 1976 there was no commercial reprocessing facility for light water reactor fuel anywhere in the world,[3] though plants for recycling fuel for other reactor types had been operated for some years in France and the UK.[4] In 1978, two reprocessing plants for light water reactor fuel were in operation,[5] one in France and another in Germany, with a total output of about 130 short tons U, corresponding to 5-6 per cent of the global spent fuel supplies in that year. In 1985, Western world reprocessing capacity, to be located in France, Belgium, Japan and Germany, is forecast at 1,400 short tons U.[6] This will permit recycling of only 20 per cent of the spent fuel elements removed from reactors in that year.[7] The forecast assumes that the US reprocessing ban will remain in force.

It is clear from the above figures that reprocessing will play an insignificant role in contributing to nuclear fuel supply, at least up to the mid-1980s. The absence of reprocessing in the US and the slowness of its development in Europe have been primarily caused by the US policy decisions discussed above. One can therefore reasonably claim that these US policies are responsible for a significant addition to forecast requirements for natural uranium.

The additional requirements due to postponed reprocessing are clearly reflected in the consecutive OECD surveys. The 1973 survey still assumes full reprocessing in the near future. The 1977 survey by contrast, assumes that there will be no reprocessing at all before 1985.[8] An approximate quantitative appraisal of the impact on requirements and demand due to the delays in recycling can be obtained in the following way. OECD's 1977 survey assesses the aggregate uranium requirements for the 1978-82 period (no recycling at all) at 267,000 short tons U_3O_8. Had recycling been introduced as expected in 1973/4, and resulted in requirements savings of 15 per cent in 1978 and 1979, and 30 per cent in 1980, 1981 and 1982, the aggregate five-year requirements figure would have been reduced to 207,000 short tons, implying a saving of

67,000 short tons. Adjusting for the more recent forecast reprocessing output in 1978-82, equivalent to 6,000 short tons of natural U_3O_8,[9] renders the total additional requirement for natural uranium due to reprocessing delays at 61,000 short tons.

This additional future demand, whose full importance emerged gradually between 1973/4 and 1977, is then the first of the external factors which contributed to the explosive market developments in uranium. Though immediate requirements for natural uranium supplies in 1974 and 1975 were hardly affected at all, the futures prices for deliveries in the late 1970s and early 1980s must have been influenced by the emergent additional demand. The impact on futures prices then spilled over to cash prices, causing the latter to rise in parallel.

The Enrichment Tails Assays

This section and the next deal with the impact on the international uranium market of government policies in the field of enrichment. To appreciate fully the significance of this issue, it should be mentioned that enrichment is a very elaborate process, involving total costs comparable to those of the natural uranium feed.

The complete dominance of the US government in this field until recently, and its fast declining role in more recent years, can be easily demonstrated. Until 1976, the US atomic authorities accounted for the entire commercial enrichment capacity in the Western world. Table 5.1 presents data for the actual situation in 1977 and plans for capacity expansion until 1990. It appears that the US share will diminish considerably during the 1980s, as the Western European enrichment units attain their full capacity.

The figures in the table underestimate the excess capacity in enrichment services, and exaggerate the US role in this activity, because a number of European utilities since about 1973 have contracted part of their enrichment requirements with the Soviet Union. The USSR enrichment services were initially offered on more flexible terms than those of the US government with regard to tails assays, timings of natural uranium delivery and changes in contracted quantities. Prices were also slightly below those in the US.[10] Germany, Finland and Spain were among the first to avail themselves of this new source of supply. Experiencing a good demand for its services, the USSR was reported early in 1975 to have raised the prices to equal those in the US.[11] The enrichment capacity available in the USSR has never been disclosed, but in 1977 enrichment exports, mainly to Western Europe, amounted to about 1 million SWUs, and they are expected to increase to between 2 and 3 million SWUs in the early 1980s.[12] So far, therefore, the USSR enrichment supply has remained marginal in relation to overall Western world needs.

Table 5.1: Western World Commercial Uranium Enrichment Capacity, and Reactor Requirements of Enrichment Services (million SWUs)

	Actual 1977	Planned 1980	1985	1990
Capacity				
US	17.0	21.0	35.4	35.4
Western Europe	0.2	6.7	17.4—21.2	18.2—31.6
Others	0	0	0.3— 7.9	5.3—22.8
Total	17.2	27.7	53.1—64.5	58.9—89.8
US share of total, percentage	99	78	67 — 55	60 — 39
Reactor Requirements (assuming 70 per cent reactor capacity factor and 0.2 per cent enrichment tails assay)	13	20	38	57

Source: Capacities from *Jahrbuch der Atomwirtschaft* (1979). Reactor requirements from T.L. Neff and H.D. Jacoby, *Nuclear Fuel Assurance, Origins, Trends and Policy Issues*, MIT Energy Laboratory Report No. MIT-EL 79-003 (MIT Press, Cambridge, Mass., February 1979), p. 81.

Two enrichment plants were established in Western Europe in the late 1970s. The first, URENCO, owned by a consortium of Dutch, German and UK interests, made its initial commercial deliveries in 1976, but its capacity has been relatively limited, and will remain below the 1 million SWUs level at least until 1982. Of greater immediate consequence is the EURODIF plant, situated in France and owned by a consortium including France, Belgium, Spain, Italy and Iran.[13] This plant became operative in 1979 with a capacity of 2.2 million SWUs, to be increased to 6.1 million in 1980 and 10.8 million in 1982.

Despite these recent developments, the US DOE is estimated to provide in 1978-9 roughly 90 per cent of total Western world commercial enrichment services.[14] The complete US monopoly of former years and the continued dominance of US supplies in this field must be kept in mind for a proper appreciation of the role of US policies regarding tails assays in this section, and of enrichment contracting in the next.

A further issue which requires clarification is the adequacy of enrichment capacity to satisfy current and future requirements, as perceived at various times. In 1971 and 1972, while the US enrichment plants were heavily under-utilised, increasing concern was expressed about the adequacy of enrichment capacity in the late 1970s and 1980s.[15] To an economist, this concern appears surprising. Given the long lead time, one would have expected that new capacity could have been installed in time to avert any future shortage. The worries must be seen against the special conditions characterising uranium enrichment: the economically unpredictable behaviour pattern of the US government, the monopolist supplier, and simultaneously the extreme barriers to entry, consisting of classified technology, indivisibilities of scale and the very large capital requirements to establish enrichment plants.[16]

The ability to expand enrichment capacity has proved far more elastic than was thought possible in the early 1970s. Contrary to the earlier expectations, this ability, along with the slower than expected growth in nuclear power generation, has created in the late 1970s a substantial glut in enrichment capacity. The figures in Table 5.1 show that this glut is expected to last well into the 1980s, and would be extended into the 1990s if all current plans to establish enrichment capacity were really accomplished.

In scrutinising the US authorities' enrichment tails policies from 1973 onwards, it is necessary to distinguish between the *operating tails assay* on the one hand and the *transaction tails assay* on the other. The operating tails assay is that at which the enrichment plant is in fact operated. The transaction tails assay is that used in the contracts between the enrichment agency and its customers to determine the amount of natural uranium to be delivered per unit of enriched uranium received. This distinction is necessary because between 1973 and 1978 the US enrichment plants applied the so-called 'split tails' policy, in which the operating assay was higher than the transaction assay. More specifically, while the transaction tails assay has in fact remained unchanged at 0.2 per cent from 1968 to the present, the operating tails assay was changed on several occasions between 1973 and 1978.[17]

The split tails policy was primarily seen as a measure to dispose of the US authorities' surplus stock of natural uranium, amounting to some 50,000 short tons, which remained from the military period. It may be recalled that the AEC had originally announced plans to dispose of this uranium through open bids.[18] In view of the weak uranium prices in the 1970-3 period, this material was in fact never offered to the market. By keeping the operating tails assay above the transaction tails assay the US authorities intended to absorb part of the surplus stock without depressing the uranium market.

While the above economic rationale for the split tails policy was valid in 1973 and early 1974, it clearly ceased to be so from 1975 onwards, when the uranium price was already very high. Inertia is probably the main reason why the split tails procedure was not discontinued until 1978. Very roughly, the policy may have absorbed some 15,000-20,000 short tons of the US government's uranium inventories.[19] These stocks could have provided some cushion at least to the price explosion, if instead they had been offered for sale, and thus added to supply.

To suggest that the split tails policy was misguided in later years does not imply that it actively contributed to the uranium market squeeze. There were no widespread expectations in 1972 and 1973 of an early release of the surplus AEC stocks to the commercial market. In the absence of a split tails policy, the authorities would probably have kept the surplus stock intact. This view is supported by the fact that the US DOE continues to hold what remains of the stock after the operating tails assay was restored to 0.2 per cent.[20] The introduction of the split tails policy in 1973, therefore, cannot be taken to have involved any reduction of the supply expected to reach the market.

Another aspect of the enrichment tails policy appears to have had a more active and manifest effect on the uranium market. In earlier paragraphs we pointed to the widespread fears in the early 1970s, both in the US and elsewhere, that enrichment capacity would become inadequate from the late 1970s onwards. One way to make existing enrichment capacity last longer was to increase the operating tails assay. As noted in Chapter 4, an increase in the tails assay would reduce the number of SWUs, but would increase the amount of natural uranium required per unit of enriched uranium. Plans to increase the tails assay featured prominently with the AEC and ERDA for several years. The split tails policy which was introduced in 1973 was seen as a temporary measure to deplete some of the surplus uranium stocks, and the authorities concerned repeatedly expressed their intention to augment the transaction tails assay in the near future to make the two equal, but this change was repeatedly postponed.[21] Early in 1975, for instance, there was an announcement of the ERDA's definite intention to raise the transaction tails assay from 0.2 per cent to 0.275 per cent effective July 1976, and to 0.3 per cent from 1981.[22] By mid-1975, the timing of the first increase had been moved forward to 1977, but in 1977 the programme was again postponed.[23] The original rationale for the planned increase in the transaction tails assay disappeared with the emergence of excess capacity in Western world enrichment plants, and the intention is unlikely to be carried out in the foreseeable future.

Though the US plans to raise the transaction tails assay were never implemented, the declared intention to do so added significantly to the demand for uranium for future delivery. That the AEC and ERDA intentions were taken seriously is illustrated, for instance, by the OECD 1973 survey, which projects uranium demand on the basis of a trans-action tails assay of 0.275 per cent from 1980, and the OECD 1975 survey which assumes the tails assay to be kept at 0.25 per cent throughout.[24] A change in the transaction tails assay from 0.2 per cent to 0.25 per cent adds to uranium requirements by about 9 per cent. Demand for uranium in the 1978-82 period, as assessed by the OECD in 1975, would have been about 20,000 short tons lower on the assumption of unchanged transaction tails assays at 0.2 per cent. One can take it that many utilities signed long-term contracts in the 1973-7 period, based on the higher future requirements figures due to increased enrichment tails. For example, Japanese utilities contracted for uranium during this period, for deliveries until 1990, on the assump-tion of a tails assay as high as at 0.37 per cent in the 1980s.[25] In this way the never-implemented US policy intention added significantly to uranium demand and hence to the price explosion which took place in this period.

The Enrichment Contracts

The third issue involving US policy to be discussed in the present chap-ter has to do with the enrichment contract terms. Until 1973 the US AEC provided enrichment services under a variety of terms, all afford-ing considerable flexibility to the buyer in terms of quantities and timing of deliveries. Beginning in 1973, the AEC introduced a new con-tracting system. From then on, its enrichment services were offered exclusively in the form of 'fixed commitment contracts'. These stipu-lated that the customer should commit himself to supply fixed quanti-ties of U_3O_8 at fixed times, and to receive fixed quantities of the enriched product over a ten-year period starting eight years after sig-nature of the contract. Thus, the customer had to make commitments stretching 18 years into the future. He was also required to make sub-stantial advance payments for the service, and was liable to heavy penalties for breaking his commitments.[26]

It is unclear what rationale lay behind this US government policy move. The simplest explanation is that the protected monopoly position permitted the US to impose contract conditions which assured convenience and ease in planning the use of its enrichment installations, without any regard for the inconvenience caused to its customers. Alternatively, the new contract policy could be seen as a pre-emptive effort to tie up enrichment demand just before the planned enrichment ventures in other countries were in a position to offer enrichment contracts.

Fearing a forthcoming shortage in enrichment capacity, utilities from the US and other countries rushed to sign the new enrichment contracts with the AEC. The eagerness to get an assurance of adequate enrichment supply actually led to an exhaustion of the US capacity. In 1974, the US authorities declared that they could not accept any further contracts, since their facilities were already fully booked. Additional contracts, conditional on the US proceeding with plutonium recycling, were signed to cover part of the unsatisfied demand.[27] In retrospect, it is clear that the prospects of a possible crash nuclear programme in the wake of the oil crisis, combined with fears about insufficient future enrichment capacity, led utilities to place enrichment orders in the US far above their expected needs as perceived at that time.[28] The fixed contract arrangement effectively locked the future enrichment volume, and hence the requirements for natural uranium, at a level above future reactor needs as perceived at that time, and very far above the future needs as perceived in the following years.

The firm decision to build the large EURODIF plant, taken late in 1973, under the impression of impending shortages of enrichment capacity, has also played an important role in this context. The participating governments in EURODIF had previously pledged themselves to taking enrichment service deliveries on a cost-recovery basis in accordance to their ownership shares, so as to utilise the project's full capacity.[29] As a result of the envisaged shortage of enrichment capacity and the difficulties in contracting US enrichment services in 1974, there was no problem in firmly committing EURODIF's entire planned capacity to the five equity holders and to Japan and Switzerland, at prices significantly above those in the US, even though some of the utilities had already signed fixed contracts to cover their needs in the US. In this way, the nervousness about future inadequacy in enrichment capacity, the consequent speedy establishment of EURODIF and the firm commitments to utilise its full capacity added substantially to the volume of natural uranium requirements effectively locked up by commitments to deliver it for the purpose of enrichment.

As a consequence of these events, there was a drastic change in uranium-contracting habits from 1973 onwards, and simultaneously a sharp upward jump in uranium demand. Utilities with heavy enrichment commitments became anxious to sign long-term uranium purchase contracts, to assure themselves of the required future supplies. The change in contracting habits is clearly evident in the US market figures. Whereas at the beginning of 1973 US utilities had contracted to buy 83,000 short tons of uranium on long-term contracts, of which only 46,000 short tons were for delivery more than three years hence,[30] at the beginning of 1976 the corresponding figures had increased to 206,000 and 160,000 short tons respectively.[31] Though corresponding figures for other countries are not available, one can conjecture that European and Japanese utilities went on a similar buying spree.

The uranium market was influenced not only by the utilities' eagerness to sign long-term contracts, but likewise by the increased potential requirements of uranium which had to be delivered to the enrichment plants, but for which the utilities had not yet contracted.

To obtain a measure of the overall distorting impact of the fixed enrichment contracts on the uranium market, one must assess the excess of actual *and* potential uranium demand due to enrichment, over the amount of uranium required to fuel existing and planned reactors. These excesses continued to be huge even after a limited relaxation introduced by ERDA in 1975, enabling its enrichment customers to alter quantities and timings in the enrichment contracts.[32]

Thus, comparing the US utilities' undertaking as of 1976 to deliver U_3O_8 for enrichment with the 1976 forecast of their need for uranium to fuel reactors means that over the five-year period 1977-81 the utilities would end up with an additional enriched uranium stock corresponding to 46,000 short tons of U_3O_8, or more than two years' current needs.[33] The situation was not much different in Europe. In 1977, European utilities found themselves with commitments to deliver uranium for enrichment in the 1979-83 period, which was about 43,000 short tons in excess of their forecast nuclear fuel requirements over the same period of time.[34] Given the sharp fall in forecast nuclear expansion which occurred between 1973 and 1977, a major share of the inventory build-up implied in these figures must clearly have been involuntary.

The underlying basis for the above distortions was of course the very substantial upside error in nuclear growth forecasts of the early 1970s, combined with the psychosis caused by the oil crisis at the end of 1973. No distortions would have arisen as a result of the fixed enrichment contracts if expectations about the future had been correct throughout the period under study. But then the erroneous nuclear forecasts had prevailed at least since 1970, without distorting the uranium market. The point made here is that the removal of flexibility in the enrichment market was instrumental in changing the uranium buying habits, in creating a bulge in demand and in freezing requirements at excessive levels, and hence responsible for a strong upward push on prices. These consequences would not have followed from the nuclear forecasting errors in isolation.

This finding concludes our analysis of how enrichment contract policies affected uranium market developments. Two digressions are in order before we close this section. The first one looks into the resolution of the huge prospective enriched uranium inventory build-up, caused by the fixed enrichment contracts. The second discusses how to approach the excessive enrichment capacity issue in an economically rational way.

While the figures presented above illustrate how enrichment contracting policies caused a tremendous surge in uranium demand after 1973, it is also important to point out that if all fixed enrichment contracts were to be enforced, there would be an explosive accumulation of enriched uranium, followed by a collapse in the demand for natural uranium as the current enrichment contracts expired.

The uranium inventory accumulation in past years reviewed in the preceding chapter has been caused to a great extent precisely by enrichment contracting policies. It appears, however, that adjustments are being found to avoid the dilemma posed in the previous paragraph. In 1978, the DOE offered some further limited relaxation of the terms in old fixed commitment contracts, and simultaneously introduced considerably greater flexibility into the new contracts that are being offered.[35] Despite heavy penalties, customers have cancelled completely numerous fixed commitment contracts with the DOE.[36] And Italy, which had been heavily over-committed to buy enrichment services, both in the US and Europe, has renegotiated its arrangement with EURODIF, thereby substantially reducing its requirements for natural uranium.[37] As a result, in September 1979, the estimated Western world excess of uranium demand due to be delivered for enrichment in 1979-83 over the quantity required to run reactors through this period had been reduced to 24,000 tons U_3O_8,[38] despite the fall in forecast nuclear capacity recorded since 1977. This is only a fraction of the excess requirements due to enrichment two to three years earlier. It is possible that the curtailment of utilities' enrichment commitments will be followed by cancellations of long-term uranium purchase contracts in some cases. In conclusion, it appears that while the inflexible enrichment contracts of the 1970s resulted in a sizeable but temporary bulge in demand, and a consequent boost in price, the potential longer-run problem of explosive inventory accumulations is now being averted.

In a booklet repeatedly quoted on earlier pages, the Uranium Institute[39] makes alternative assessments of future demand for natural uranium. One of these is based on current and forecast nuclear reactor requirements. Another assumes that uranium demand will be at the much higher level determined by full capacity utilisation in existing and planned enrichment installations. The authors admit that the two forecasts constitute extremes, that all planned enrichment plants may not be built and that demand for uranium will in fact settle somewhere in between. However, they also note that at present and at least up to the mid-1980s the owners of enrichment plants are in a strong enough bargaining position, despite the excess enrichment capacity, to assure that their interest in full utilisation of this capacity will be achieved.[40] The ongoing cancellation of enrichment contracts and the consequent shrinkage of excess uranium demand discussed above hardly support this view. Be that as it may, there is an interesting and important issue which requires clarification. Given that the enrichment capacity is higher than that warranted by current nuclear reactor requirements, what are the conditions under which it would be rational from the global economic point of view to keep the excess enrichment capacity fully occupied by production of enriched uranium in excess of current reactor needs?

One way to approach this problem is to assume that all uranium production and uranium enrichment are handled by one single firm. To simplify the problem further, we take it that this firm can dispose of all its enriched uranium production at full cost coverage (including normal profits) up to the level required to satisfy current reactor needs, but that it can dispose of nothing beyond that volume. We also assume that mining and enrichment capacity can be established instantaneously, that marginal costs up to the capacity limit are constant, and that the firm knows, but has no control over, the rate of growth in nuclear capacity and the concomitant uranium fuel needs. Furthermore, we assume that there is no inflation.

In a firm facing these circumstances it will be economically rational to employ the excess capacity in enrichment to the extent that the total cost of natural uranium plus the variable cost of enrichment plus the cost of storage of the enriched uranium until needed is lower than the total cost of natural uranium plus the total cost of enrichment. The numerical example presented below elaborates on this statement.

As a point of departure, let us take it that natural uranium is enriched to a 3 per cent content of U_{235}, with a 0.2 per cent tails assay. This gives us the following approximate relationship:[41]

$$1 \text{ lb } U_3O_8 + 0.3 \text{ SWU} = 0.15 \text{ lb } U_{enriched}$$

The full cost (as distinct from price) of natural uranium in our example is taken at $30. The full cost of enrichment is taken at $100 per SWU.[42] Hence, the total cost to convert 1 lb of U_3O_8 to 0.15 lb of enriched uranium works out at $60.

Once enrichment capacity has been established, the fixed enrichment costs can be disregarded. Operation of this capacity will be economical so long as it provides for some contribution towards fixed costs over and above the recovery of variable costs. To complete our example of the firm with excess capacity in enrichment, therefore, we need to know the proportions of variable and fixed costs in enrichment. The DOE provides a split-up of the enrichment cost structure, indicating that of the total cost of just about $100/SWU, $45.13 constitutes the electric power cost and $9.47 other operating costs.[43] Without any attempt to be precise, we take the figures to suggest that roughly one-half of the total enrichment costs are variable.[44] The firm with excess capacity in enrichment will find it economical to use this capacity only if it can recover at least the full natural uranium cost of $30/lb U_3O_8 (since by assumption there is no excess capacity in natural uranium production), and the variable enrichment cost of 0.3 SWU, amounting to $15, or a total of $45 per 0.15 lb $U_{enriched}$.

With our very particular assumption about the enriched uranium market, there are no means at all to dispose immediately of the excess enriched uranium which is produced. Hence, such uranium would have to be stored until nuclear reactor requirements had grown above the existing enrichment capacity. At that time the inventory can be disposed of at the full price of $60 per 0.15 lb $U_{enriched}$. To decide whether or not to produce now in excess of current requirements, the firm must add to the cost of $45, computed above, the cost of carrying the stock until the sale can be made. Excess production will be economical only if the total cost remains below $60. With an overall inventory cost at 10 per cent per year (including a rate of interest on the capital tied up, concomitant with our assumption of no inflation), production in excess of requirements is worth while if the enriched uranium can be disposed of within three years.[45]

Though our example is based on several highly simplified assumptions, it does contain all the major elements required to assess the global economics of operating excess enrichment capacity. It also permits us to explore how these economics change as we alter some of the parameters used above. Thus, *ceteris paribus*, the scope for economic utilisation of excess enrichment capacity will be reduced as a consequence of:

an increase in the natural uranium costs;
an increase in the proportion of variable costs in uranium enrichment;
an increase in the tails assay;
an increase in the cost of carrying inventories; and
a slow-down in nuclear growth forecasts.

The probability that an economically rational approach to this issue on a global scale will be adopted is of course small. Even though public authorities concerned with social rather than private goals are heavily involved in all stages of nuclear fuel production and use, the governmental decision-makers are likely to aim at maximising national rather than global objectives. Market position and bargaining power will be used by countries with export capacity in enrichment to have that capacity fully used, even if the end is economically irrational on a global scale. Countries importing enrichment services are likely to adopt a similar attitude and try to avoid enrichment imports which appear uneconomical from the national standpoint. The degree of and terms for the utilisation of the excessive enrichment capacity will therefore probably be determined through an oligopoly-oligopsony bargaining process, whose end result will depend on the parties' economic, financial and political strengths and vulnerabilities rather than on a rational assessment of what is globally economical.

A Summary of Conclusions

Our analysis has shown how policy decisions, taken predominantly in the US, with regard to (a) the reprocessing of spent uranium fuel, (b) enrichment tails assays and (c) enrichment contracting, added substantially to uranium demand after 1973. We have attempted to quantify the impact on the uranium market by assessing the additional five-year requirements for natural uranium resulting from these policies.

Thus, we estimate that the gradual deferment of the introduction of reprocessing and plutonium use added to the 1978-82 uranium demand by more than 60,000 short tons. The 1975 policy intentions to raise the transaction tails assay added a further 20,000 short tons to the prospective uranium demand in the 1978-82 period. In the US in 1976, uranium requirements for enrichment from 1977 to 1981 exceeded by 46,000 short tons the amount needed to fuel existing and planned reactors. And in Europe in 1977, the corresponding figure for the 1979-83 period amounted to 43,000 short tons. We have not been able to quantify the Japanese excess demand for uranium due to enrichment commitments.

These additional actual and prospective demand figures resulting from government policy are indeed considerable. One must be cautious about aggregation, first, because some double counting may be involved and, second, because the figures relate to somewhat different periods. Even if precision is lacking, it suffices to note that the total addition to demand, corresponding to more than 30,000 short tons per year, is of the same order of magnitude as world production in the mid-1970s. The impact of these policy decisions on the uranium market from 1973 onwards must therefore have been quite substantial.

The bulge in additional actual and potential demand, due to the above policies, coincides with the 1974-6 period of fast-rising uranium prices. It remains to note that only the first of the three policies, i.e. the one to defer reprocessing, continues to exert its full influence on uranium demand. It is reasonably clear that the DOE enrichment tails assay will not be increased, and we have shown in the body of the chapter the very substantial evasion by utilities in 1978 and 1979 of excessive enrichment contract engagements. The reduction in future uranium demand caused by these recent developments is superimposed on the reduction in future requirements due to slower nuclear growth forecasts. The combined effect of these depressive forces certainly provides an important explanation to the sharp fall in real uranium prices in 1978 and 1979.

A number of subsidiary issues, supporting or otherwise connected with the above major themes, have also been pursued. For instance, we have analysed the withering of US dominance in the field of enrichment. We have studied the misguided US policy to absorb surplus uranium stocks from the military period through a split tails enrichment policy. And we have tried to clarify the conditions under which it would be economically rational to enrich uranium in excess of current nuclear reactor needs.

Notes

1. *Uranium Price Formation*, EPRI (October 1977), prepared by Charles Rivers, Cambridge, Mass., p. 4.14.

2. *Nukem Market Review* (April 1977).

3. J.S. Nye, 'Balancing Non-proliferation and Energy Security', *Uranium Supply and Demand* (The Uranium Institute, London, 1978), p. 327.

4. OECD, *Uranium Resources, Production and Demand* (OECD, Paris, December 1975), p. 63.

5. *Nukem Market Review*, no. 8 (1978), p. 6.

6. *Atomwirtschaft* (February 1979). Reprocessing capacity is measured in terms of the uranium content of the spent fuel elements. This uranium has a higher concentration of U_{235} than natural uranium. The quantities of uranium obtained from reprocessing have therefore to be adjusted upwards to get their equivalent in terms of natural uranium.

7. *The Balance of Supply and Demand 1978-1990* (The Uranium Institute, London, 1979).

8. OECD, *Uranium Resources, Production and Demand* (OECD, Paris, August 1973); ibid. (December 1977).

9. *The Balance of Supply and Demand 1978-1990*.

10. *Nukem Market Review* (March 1973).

11. *Nukem Market Review* (January 1975).

12. *Jahrbuch der Atomwirtschaft* (1979).

13. In 1979, Iran was considering the possibility of leaving EURODIF, but was hesitant to do so in view of the severe economic consequences such a step would involve. See *Nuclear Fuel* (25 June 1979).

14. T.L. Neff and H.D. Jacoby, *Nuclear Fuel Assurance, Origins, Trends and Policy Issues*, MIT Energy Laboratory Report No. MIT-EL 79-003 (MIT Press, Cambridge, Mass., February 1979).

15. *Uranium Price Formation*, pp. 3.37-8.

16. In 1975 it was estimated that the minimum economic capacity of a standard technology diffusion enrichment plant was close to 10 million SWUs, sufficient to serve 75 light water reactors of 1,000 MWe each. The investments to establish such a plant were assessed at $2.5 billion. See *Atomwirtschaft* (September 1975), p. 413.

17. *Uranium Price Formation* and several issues of *Nukem Market Review*. The operating tails assay was increased from 0.2 per cent to 0.275 per cent in 1973, and to 0.3 per cent in 1974, but was lowered to 0.25 per cent in 1977, and returned to the original value of 0.2 per cent in 1978.

18. See Chapter 3.

19. Taking the overall natural uranium deliveries for enrichment between 1973 and 1978 at 125,000 tons, and assuming the average operating tails assay in the five-year period to have been at 0.275 per cent, renders an additional uranium requirement of 17,000 tons to meet the contractual enrichment obligations based on a transaction tails assay of 0.2 per cent.

20. See, for instance, *Uranium Price Formation* or OECD (December 1977).

21. *Uranium Price Formation*, p. 3.39.

22. *Nukem Market Review* no. 1 (1975).

23. *Uranium Price Formation*, p. 3.39.

24. OECD (August 1973), p. 110. OECD (December 1975), p. 32.

25. Neff and Jacoby, *Nuclear Fuel Assurance*, p. 90.

26. *Uranium Price Formation*, pp. 3.41 and 3.42. Also see F. Oboussier, 'Die Versorgung der Bundesrepublik Deutschland mit angerichtertem Uran', *Glückauf*, no. 9 (1976).

27. F. Oboussier, 'Die Versorgung'.

28. P. Jelenek Flink, 'The Impact of Enrichment Policies on the Uranium Market', *Uranium Supply and Demand* (1978).

29. Jelenek Flink, 'Impact of Enrichment Policies', p. 55.

30. J.H. Lorie and C.S. Gody, 'Economic Analysis of Uranium Prices', (9 July 1975), a report prepared for Westinghouse Electric Corporation.

31. J.A. Patterson, 'Uranium Market Activities' (US ERDA, Washington, DC, October 1976).

32. *Uranium Price Formation*, p. 3.42.

33. *Uranium Price Formation*, Chapter 4. Uranium requirements have been approximated on the basis of linear growth from official forecasts providing annual requirements with five-year intervals.

34. *Nukem Market Review* (November 1977).

35. *Nukem Market Review* (July 1978).

36. *Nuclear Fuel* (20 August 1979).

37. *Nuclear Fuel* (29 October 1979). The arrangement involved an Italian purchase of uranium tails in the US, containing 0.3 per cent U_{235}. These tails will be enriched by EURODIF to 0.7 per cent U_{235}, i.e. equal to natural uranium. The arrangement reduces Italian requirements of natural uranium on two counts: first, because no natural uranium material is needed to fulfil the EURODIF contract and, second, because the material received from EURODIF is equivalent to natural uranium. Though in itself the arrangement is unlikely to be economical, it apparently offers Italy a better alternative than that of cancelling the EURODIF contract and paying the fines.

38. *Nukem Market Review*, no. 9 (1979).

39. *The Balance of Supply and Demand 1978-1990*.

40. P. 11.

41. We can assume that the cost of conversion is included in the cost of natural uranium.

42. The US DOE charged about \$100/SWU in the latter half of 1979. See *Nukem Market Review*, no. 9 (1979).

43. *Nukem Market Review*, no. 9 (1979).

44. In cases where the electric power capacity has been established for the exclusive use of the enrichment plant, one could argue that it is a fixed cost, since it cannot easily be diverted to other uses.

45. We disregard the saving in capital cost resulting from inventory build-up, due to the deferment of enrichment capacity expansion through the period of inventory disposal.

6 THE IMPACT OF THE OIL CRISIS

Introduction

Petroleum dominates the energy consumption pattern in the OECD area. In 1972, it accounted for more than 55 per cent of overall energy use, while the nuclear share was no more than 1 per cent.[1] Even in 1978, after five years of high petroleum prices, this fuel still accounted for more than half of overall energy use, but the nuclear share had increased to 3 per cent of the OECD total.[2] So long as its overwhelming dominance remains, developments in oil will have a profound impact on the markets for all other energy materials.

Our price analysis of uranium, petroleum and coal in Chapter 2 supports the above thesis. Average nominal petroleum prices rose more than fourfold between 1973 and 1975. The prices for coal and uranium followed upwards in sympathy, with a lag. Occasionally it has even been suggested that the petroleum price changes in themselves suffice to explain the entire price performance in the uranium market after 1973. While we agree that the petroleum events had a very strong impact on uranium prices, it is our opinion that several other factors also played important roles in determining uranium price developments in the mid-1970s. Some of these have already been dealt with in earlier pages. Others will be analysed in the following chapters.

The present chapter is devoted to clarifying the interrelationships between the oil market events and uranium prices. Three connections appear to be particularly important. The first one can be disposed of briefly. The tripling of the petroleum price towards the end of 1973 and the simultaneous export embargoes against certain oil-importing countries implemented by major Arab oil producers created a high degree of confusion and insecurity about international energy trade in general. One can reasonably conjecture that this insecurity strengthened the utilities' desire to sign long-term purchase contracts, and increased their propensity to hold inventories. The ensuing additional uranium demand occurred mainly during 1974, and thus coincided with the utilities' buying spree in conjunction with the change in the US enrichment policies, as described in the previous chapter. The two factors reinforced each other in turning the uranium market in that year into a sellers' market.

The remaining relationships are far more complex. The second one has to do with the economics of substitution, while the third involves appropriation of the quasi rents resulting from higher energy prices. The distinction between these two relationships can perhaps be clarified with the help of Chart 6.1. Current prices of final energy (for instance electricity) are determined by the price of oil, the dominant energy material, at p_1. S is the long-run supply (marginal cost) curve for nuclear electricity,[3] and Q_1 is the current nuclear production level. Suppose that events in the oil market increase the price of final energy to p_2. Nuclear power production will now expand towards Q_2 by substituting for part of the oil-based energy. As the share of nuclear in total energy output is expanded, the overall nuclear supply cost as well as the supply costs of its various components (including uranium) will tend to rise. Substitution will continue until the overall marginal supply cost equals the new price.

Substitution in the energy field is a slow and drawn-out process, in part because of the long gestation periods of investment planning and execution. A temporary quasi rent, equal to $p_2 p_1$ per unit of energy, will emerge while nuclear power production remains at Q_1. This rent may become permanent if substitution of nuclear for other energy forms is prevented, for instance by public regulation or by uncompetitive market conditions. When substitution is temporarily or permanently hindered, there will be tendencies to appropriate the rent by the nuclear utilities, as well as by their input suppliers, including uranium producers.

The rest of the present chapter is devoted to clarifying the processes of substitution and quasi rent appropriation in nuclear power production, consequent upon the oil price rise of 1973/4.

Numerous data on costs at different times in alternative energy production processes are presented as we go along. It should be clear that cost estimates are bound to vary considerably, depending on underlying assumptions regarding, for instance, longevity of installations and interest rates, or because environmental and safety precautions and accounting practices may differ both over time and between countries. Our intention in presenting the data is not to provide precise comparisons of costs and prices over time and between energy production alternatives, but merely to illuminate within broad ranges the economic consequences of the 1973/4 petroleum price increase for nuclear power generation, and specifically for uranium production.

Chart 6.1: The Process of Substitution and Quasi Rent Appropriation in Nuclear Power Production

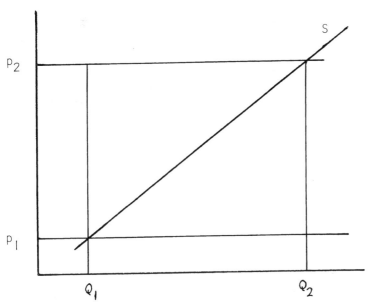

The Economics of Substitution

When competitive conditions prevail, economic theory postulates that as the price of a product is raised, consumers will gradually shift their demand to substitutes, until a new equilibrium is reached with a reduced market share for the more expensive product, and increased market shares and prices for its substitutes. The extent of the shift following a relative price change will depend on how well existing substitutes can satisfy the functions performed by the product whose price rose initially, and on the price increase required to induce additional substitute supplies.[4] Before discussing the more complex petroleum/nuclear substitution relationship, a very simple case of copper is presented to illustrate the above argument.

If Zambia unilaterally raised its copper price, the likely effect would be a complete substitution away from Zambian copper in favour of other suppliers. This is, first, because copper produced elsewhere is a perfect substitute for that of Zambian origin and, second, because given a period of adjustment, the rest of the world could easily replace Zambia's limited share of world supplies without a noticeable rise in costs and prices.

Substitution would only be partial, on the other hand, if all primary copper producers jointly raised the copper price. The supplies of the (almost) perfect substitute of old copper scrap account for a limited proportion of overall supply, and could not possibly be raised to substitute for the overall primary supply. Even a partial replacement of primary copper would involve higher scrap costs and prices due to the more intensive collection and recycling requirements. The rising scrap price in turn would reduce the inducement to substitution. Neither could substitution of alternative materials like aluminium for copper be more than partial. This is because aluminium is a far from perfect copper substitute. The relative increase in the price of primary copper would lead to replacement by aluminium, but only in uses where the two metals' functions were reasonably overlapping.

The extent of substitution following a relative price change would also vary with time. The ability to substitute will increase over time, as the capacity to supply the substitute materials is expanded, and as the consumers adjust their habits or their capital equipment to the use of the alternative product.

Commercially traded uranium has been used so far almost exclusively for electricity generation,[5] and nuclear electricity is a perfect substitute for electricity from petroleum- or coal-fired power stations. In a long-run equilibrium situation one would therefore expect the total costs of marginal units to be the same, irrespective of whether petroleum, coal or uranium was used as fuel. If the cost of petroleum- and coal-based electricity rose in relation to that of nuclear, the consequence would be an acceleration in the growth of nuclear electricity, and an expansion of its market share, until its marginal cost had risen to equal that of electricity from petroleum and coal, or until it had substituted completely for petroleum and coal in electricity generation. In view of the limited nuclear share in total electricity production, the latter is unlikely to occur in the foreseeable future. Nuclear electricity costs would probably have to rise very substantially, if nuclear expanded to cover all electricity demand in this century. In this, the nuclear versus petroleum substitutability conditions resemble those of old copper scrap versus primary copper.

Nuclear electricity is also an imperfect substitute for petroleum and coal in a variety of other energy uses. As the relative price of the fossil materials increased, nuclear electricity could, for example, substitute for them in some cases in the production of heat. Here, the nuclear versus petroleum substitutability conditions resemble those of aluminium versus primary copper.

A suitable starting-point for the discussion of possible substitution impacts in the wake of the oil price increase may be a rough indication of the total electrical generation costs in new nuclear, and fossil fuel, plants as assessed in 1973. The figures provided in Table 6.1 are based on US data,[6] and assume fuel prices at $6.50/lb U_3O_8, $8.50 per short ton of coal[7] and $3.30 per barrel of oil. The figures suggest that total costs of nuclear electricity for base load application were 10-15 per cent below fossil fuel electricity. (The nuclear advantage would be less in intermediate load applications.) A French assessment of 1972 confirms this finding.[8] After adjusting the petroleum price for the 15 per cent increase in real terms which occurred between 1972 and 1973, the French figures, like the American ones, suggest that in 1973 nuclear electricity from new power plants would have been about 10 per cent cheaper than electricity based on petroleum.

Table 6.1: Estimated Generation Costs for New 1,000 MWe Steam-electric Power Plants at 1973 Costs and Prices (US cents per KWh)

	Nuclear	Coal	Oil
Capital	0.79	0.74	0.54
Fuel	0.23	0.37	0.55
Operating	0.07	0.11	0.05
Total	1.03	1.22	1.14

Electricity prices charged to large industrial users in 1973 were 1.25 US cents per KWh in the US,[9] but higher in most other industrialised countries.[10] It would appear, therefore, that all three forms of electricity generation could recover their full costs in 1973.

The low relative cost levels of nuclear electricity were seen in the early 1970s to result in a very substantial substitution in the long run of nuclear for coal- and oil-fired power stations. This appears clearly, for instance, in the 1973 OECD forecast of nuclear power expansion,[11] which is seen to proceed at much higher rates than overall electricity demand. In a more precise assessment using energy price and cost data just before the 1973 oil price increase, the OECD Secretariat projects an increase in nuclear power's share of total electricity generation in the OECD area from 9 per cent in 1972 to 66 per cent in 1985. The same report also envisages the share of electricity in overall energy use to rise from 12 to 15 per cent over the corresponding period.[12] Implicit in this rise in the electricity share is some substitution of nuclear electricity for non-electrical uses of petroleum and coal. These were then the substitution forces in operation before the oil price rise.

By 1975, the economic competitiveness of nuclear electricity had been considerably improved as a result of the petroleum, coal and uranium price changes which occurred over the two years. The reason for this improvement is primarily due to the fact that coal and petroleum accounted for as much as 30 and 50 per cent respectively of total electricity generating costs, while uranium oxide constituted only a fifth of the nuclear fuel cost and no more than 4-7 per cent of the overall electricity generation cost in 1973.[13] With these cost relationships, all other things being equal, the competitive position of nuclear electricity is bound to improve even if uranium prices rise three times as fast as the prices of the two fossil fuels.

Table 6.2 is a very rough attempt to describe the new base load cost relationships between the three alternative modes of electricity generation in new power plants at the end of 1975. The figures in Table 6.1 constitute a starting-point for the determination of the new data. In addition, the following assumptions have been used.

(1) Operating costs in all three power generation systems have been increased in parallel with inflation between 1973 and 1975, i.e. by 25-30 per cent.
(2) Capital costs in all three power generation systems have been increased by a total of 45 per cent, to account for 25-30 per cent inflation and a 15-20 per cent increase in fixed dollar investment costs. (An important reason for the increase in real investment costs is undoubtedly the increasing environmental and safety regulations imposed on power plant designs in numerous countries in the 1970s.)[14]
(3) The doubling of nuclear fuel costs involves an increase in the price of U_3O_8 from $6.50/lb to $40, as well as estimates of actual increases in other nuclear fuel costs,[15] such as conversion, enrichment, fabrication, treatment of spent fuel and inventory charges. Coal and petroleum prices are taken to have increased by factors of 2.6 and 4.1 respectively, i.e. to $22 per short ton of coal and to $13.50 per barrel of oil.

The estimated generation costs[16] can again be compared with electricity prices charged to large industrial users in 1975. These were 2.07 US cents per KWh in the US, and higher in most other industrialised countries. Thus, in the US at least, electricity prices were inadequate to cover the full generation costs of new coal- and oil-fired power stations, as assessed in the table.

Table 6.2: Estimated Generation Costs for New 1,000 MWe Steam-electric Power Plants at 1975 Costs and Prices (US cents per KWh)

	Nuclear	Coal	Oil
Capital	1.15	1.07	0.78
Fuel	0.50	0.96	2.26
Operating	0.09	0.14	0.06
Total	1.74	2.17	3.10

One would normally have expected the sharply improved competitive edge of nuclear electricity to strengthen the forces of substitution with a consequent increase in the forecast rate of expansion of nuclear power. In fact, neither of these tendencies has occurred. The rate of substitution in favour of nuclear has not increased its pace. This may be due to the fact that the pre-oil-crisis rate of substitution was already at the maximum of what was technically feasible. Also, the opposition to nuclear power and the ensuing emergence of regulatory constraints have worked against the relative competitive advantage gained by nuclear electricity generation between 1973 and 1975. Furthermore, there have been sharp downward revisions in forecast overall world energy requirements, following the energy price rises of 1973-5, which also resulted in reduced expansion plans for the nuclear sector.

Since neither the rate of substitution nor the overall expansion plans for nuclear energy was increased between 1973 and 1975, it would be difficult to claim that the oil price changes in the early 1970s had an unequivocal upward impact on the current or expected future marginal cost of nuclear power generation. This conclusion has to be somewhat guarded, however, since the emergent rates of substitution and of growth in nuclear electricity are the result not only of the oil price change but also of several other factors which were simultaneously at work.

The Appropriation of Quasi Rents

The general upward price push in all types of final energy caused by the oil price increase led to the emergence of quasi rents in energy production systems based on raw materials other than oil. This is clearly apparent in the data in Tables 6.1 and 6.2 The increase in nuclear fuel and coal prices between 1973 and 1975 can be seen as a result of the appropriation by the producers of these fuels of part of this quasi rent. The increase in the difference between electricity prices and total nuclear generation costs (0.22 cents per KWh in 1973 versus 0.33 cents in 1975, using our cost data and US electricity prices) constitutes the quasi rent absorption by the electrical utilities.[17] The quasi rent absorption described here need not necessarily require oligopolistic bargaining strength. Under the circumstances which evolved, some of the benefits may have fallen automatically into the hands of price-taking input suppliers.

Over time one would expect the overall costs of the three alternative modes of electricity generation to be equalised through further quasi rent appropriation. In the short run, however, large differences can remain. The figures in Tables 6.2 amply illustrate that this was the case in 1975. The major reason why the nuclear and coal fuel producers did not increase their prices even more than they did was that existing oil-fuelled power stations would continue to run even if they did not get their fixed costs covered. Given the slack in electricity demand in 1975, caused both by the world depression and by the increased electricity price, assurance of full capacity utilisation in nuclear- and coal-based electricity required fuel prices at which total generating costs were lower than the variable costs of oil-based electrical power. The uncertainty as to whether the OPEC cartel would be able to keep petroleum prices high remained for several years after the oil crisis, and was another factor constraining the price increases in nuclear fuels and coal. The ability of alternative fuel producers to increase prices was also constrained by the oligopsonistic market structure among utilities in most countries, and the consequent bargaining power that such utilities could exert in fuel supply negotiations. Finally, nuclear fuel producers were anxious to ascertain that their pricing policy allowed utilities a higher relative profit margin in nuclear as compared to coal- and oil-based electricity generation, in order to assure continued growth of nuclear's market share.

Our data base for the analyses pursued in this chapter is indeed shaky, and we underline again that the figures presented should be regarded as illustrations of orders of magnitude or of the directions of change rather than precise and unambiguous representations of exact cost conditions. Given this weak data foundation, it is impossible to quantify the magnitude of freedom afforded to nuclear fuel producers to raise their prices following the oil price change. Neither do we know how efficient these fuel producers were in appropriating the emergent quasi rent. It is therefore impossible to establish in an unequivocal way the impact of the petroleum price increase on uranium prices.

One way to obtain at least a feel for the magnitudes involved could be to proceed as follows. Let us assume that the overall nuclear fuel cost increase which occurred after 1973 provides a fair representation of the overall quasi rent resulting from the oil price change appropriable by the nuclear fuel producers. In the absence of any other change, one may hypothesise that the distribution of this quasi rent among those involved in the various stages of nuclear fuel production could be in proportion to the share of each stage in total nuclear fuel cost before the oil price increase.

Our attempt to carry out this exercise extends in time from 1973 until 1977, to allow for any lag in the nuclear fuel producers' efforts to appropriate the quasi rents after 1973/4. Also, we concentrate attention on the production of $U_3 O_8$, conversion, enrichment and fabrication stages of nuclear fuel production. To get a clear-cut view of the issue in focus we abstract from other elements of the total nuclear fuel costs, because we feel that they might blur the picture. This would be true, for instance, of the spent fuel costs or credits for recycled uranium and plutonium. The very sharp changes in these cost and credit items which occurred in the early and mid-1970s were due to public policy changes unconnected with the oil price increases.

Table 6.3 shows that the total cost for the four steps in nuclear fuel production under scrutiny increased from 0.138 US cents per KWh in 1973 to 0.412 cents in 1977, i.e. by a factor of 3 in nominal and by a factor of 2.2 in real dollar terms. The share of $U_3 O_8$ in the total cost in 1973 was 25 per cent. An unchanged share for uranium in the total cost in 1977 would have been attained if the $U_3 O_8$ price had risen to $19.10 in 1977 in nominal dollars.

The result of our exercise is that in a world unaffected by other factors, the increase in oil prices might have caused the uranium price to rise from $6.40 in 1973 to $19.10 in 1977 in current dollar values, or from $8.10/lb to $17.30 in fixed (1975/6) dollars.

The world was not unaffected by other factors, however, and the uranium price rose far beyond the level which our exercise suggests it might have reached on account of the higher oil prices. The preceding chapter discussed some of the developments, apart from the petroleum price, which affected uranium market conditions during this period. The fact that the uranium price rose faster than the prices of conversion, enrichment and fabrication suggests yet another factor at work. There is a possibility that organisational changes on the supply side of the uranium market in the early 1970s improved the bargaining power of uranium producers as compared to the bargaining power of the producers of conversion, enrichment and fabrication services, thereby enabling them to appropriate a disproportional share of the quasi rent created by the oil price rise. We propose to explore the validity of this contention in the following chapter, and to study the implications for the uranium market of the uranium producers' cartel which had its antecedents in the depressed conditions of 1970 and 1971, but which became really consequential only in 1974.

Table 6.3: Changes in Prices and Costs of Certain Components of Nuclear Fuel

	1973		1975		1977	
	Price	Cost, US Cents/ KWh	Price	Cost, US Cents/ KWh	Price	Cost, US Cents/ KWh
U_3O_8	$6.40/lb	0.035	$23.70/lb	0.128	$42.20/lb	0.228
Conversion	$1.35/lb U	0.007	$1.80/lb U	0.009	$2.00/lb U	0.010
Enrichment	$35/SWU	0.063	$50/SWU	0.091	$70/SWU	0.127
Fabrication	$70/kg U	0.033	$98/kg U	0.046	$100/kg U	0.047
Total of 4 items		0.138		0.274		0.412

Note: A variety of sources have been used to compile the above information. They include *The Nuclear Industry*, US AEC, Wash. 1174, (1971-4); *Monthly Energy Review*, US DOE (July 1976 and July 1979), *NUEXCO Market Report* and *Nukem Market Review* for the three years quoted. The prices given reflect representative price levels in the US market.

Notes

1. OECD, *Energy Prospects to 1985* (Paris, 1974), vol. 1, p. 57.
2. BP, *Statistical Review of the World Oil Industry* (London, 1978).
3. We use the standard textbook assumption of an upward slope in the long-run supply curve.
4. A formal treatment of the issues is found in, for instance, C. Van Duyne, 'Commodity Cartels and the Theory of Derived Demand', *Kyklos*, vol. 28 (1975).
5. To keep the substitution issue at a simple level, we disregard the limited use of nuclear reactors for direct generation of heat.
6. *The Nuclear Industry 1973*, US AEC, Wash. 1174-73, p. 15.

7. This is the US price. Coal prices in other major countries like France, Germany and Japan were substantially higher than in the US throughout the 1970s. See OECD, *Energy Statistics 1975/1977* (OECD, Paris, 1979).

8. *Techniques de l'Energie* (June 1979), p. 25.

9. *Monthly Energy Review* (US Dept. of Energy) (July 1979).

10. OECD, *Energy Statistics 1975/1977*. Electricity prices charged to large industrial users in some countries are given as follows:

	France	Germany	Japan	Sweden
1973 price in US cents/KWh	1.46	2.37	1.44	1.03
1975 price in US cents/KWh	2.27	3.46	2.76	1.88
Change in nominal price (per cent)	55	46	91	83

For comparison, the US price to industrial users rose from 1.25 cents/KWh in 1973 to 2.07 US cents in 1975, or by 66 per cent. (See *Monthly Energy Review* (July 1979).)

11. See Chart 4.4

12. OECD, *Energy Prospects to 1985* (OECD, Paris, 1974), vol. 1, p. 57.

13. General Electric, 'Uranium 1973-1985, Materials Resources Planning Report' (December 1973), p. V-2; E. Svenke, 'Uran — globala aspekter på råvaruförsörjningen för kärnkraft', lecture delivered at the Swedish Geological Society symposium, 28 March 1974, and Uranium Price Formation, EPRI (October 1977), prepared by Charles Rivers, Cambridge, Mass., p. 4-3.

14. For evidence of the fact that capital costs developed in parallel in nuclear and fossil fuel power stations, see for instance *Jahrbuch der Atomwirt-schaft* (1977), p. B 43, or *Technique de l'Energie* (June 1979), p. 26. Both sources note that the investment costs increased faster than inflation. *Technique de l'Energie* puts the capital cost increase over the six-year period 1972-8 in constant money at 40-50 per cent.

15. Uranium prices did not reach $40 until early 1976. The development of costs and prices for other major components in nuclear fuel production are treated in greater detail in the following section.

16. One may compare our figures in Table 6.2 with a French assessment of total generation costs in new power plants in 1978 dollars, which work out at 2.36 US cents for nuclear, 2.86 cents for coal and 3.20 cents for petroleum. See G. de Carmoy, 'Nuclear Energy in France', *Energy Economics* (July 1979). Another study puts the Swedish nuclear generation costs in new plants in 1978 dollars at 2.15 cents, and states that this is far below the total coal- and petroleum-based electricity generation costs in that country. See O. Vesterhaugh and B. Blomsnes, 'Trends in Nuclear Power Costs in Sweden', *Nuclear Engineering International* (December 1979).

17. One would also have expected part of the quasi rent to accrue to the suppliers of capital equipment. The fact that the investment costs of nuclear and coal power plants did not rise faster than the investment costs of petroleum power plants suggests that the producers of capital equipment used in nuclear and coal power stations were not very successful in appropriating the quasi rent.

7 THE INTERNATIONAL URANIUM CARTEL

Introduction

Until now most of our analysis has been based on the presumption that reasonably competitive conditions characterised the uranium industry. This seems to have been the case at least until 1972, when uranium-producing companies, along with producing country governments, are reported to have initiated an organised collaboration on price and supply policies. This collaboration did not lead to any significant economic impact until 1974. By that time, however, the trust and cohesion established among producers had greatly increased their bargaining power and permitted them to turn the emergent events in the uranium market to their own advantage.

In the section which follows we study the necessary preconditions for successful price-raising cartel action, and try to determine the extent to which such preconditions prevailed in the uranium market. We thereafter analyse the actual operations of the cartel. This analysis is based on established facts for part of the period under scrutiny and on reasonable inferences, analogies and conjectures when factual information is not available.

The Preconditions for Successful Cartel Action

The necessary (but not always sufficient) preconditions for successful cartel action can suitably be treated in two steps. In the first one we explore the structural circumstances which have to be present for collusion among sellers to be feasible. In the second we clarify the conditions under which collusion with regard to prices or quantities transacted can render profitable results to the participants. Obviously, where profitable results cannot be attained, collusion is unlikely to develop even if the structural circumstances for its emergence are favourable.

The second step covers ground related to that which was explored in the preceding chapter's discussion of the economics of substitution. That discussion dealt with the conditions under which nuclear electricity can be substituted for electricity from other sources. In what follows we discuss, instead, the conditions under which consumers can substitute away from a uranium cartel's supply. Perhaps the most important structural preconditions for collusion among sellers are that the market should be concentrated, i.e. that relatively few sellers dominate overall supply, that their respective market shares are not too dissimilar, and that the transactions around which collusion revolves (price, sales volume) can be controlled and verified.[1]

The uranium industry certainly satisfies the above conditions. Relatively few centres control a very large share of overall supply. National production shares may be a relevant measure for strategic raw materials in view of the frequent government involvement in such markets. Viewed this way, uranium production is exceedingly concentrated, with four countries accounting for 90 per cent of the global total. In Chart 7.1 we compare this measure of concentration for uranium with that for six other mineral commodities. Uranium comes out on top among the seven and far above petroleum. As appears from the chart, only nickel attains a degree of concentration comparable to that of uranium.

Though national concentration is particularly important for uranium, in view of the profound governmental intervention and control in the production and trade of this commodity it is also important to study concentration among corporate decision units. This, too, appears to be very high in uranium. Virtually all South African output is controlled by Nufcor, the joint marketing company of the gold/uranium mines.[2] The French government's 'Commissariat à l'energie atomique' (CEA) owns about nine-tenths of the uranium production installations in France, and has heavy participation in the mines of Gabon and Niger.[3] Through Uranex, the export marketing agency for French, Gabonese and Niger uranium,[4] CEA can be seen as virtually in control of the entire supply from the three countries. RTZ controls a large and widespread empire of uranium interests. And the biggest US and Canadian uranium producers account for sizeable shares of these countries' overall output. The information compiled in Table 7.1 indicates that six decision units control as much as two-thirds of current world production capability.

The degree of concentration among corporate units and the variation in market shares among leading uranium producers is compared in Table 7.2 with conditions in a few other minerals industries. Even though uranium does not come out on top among the six industries studied, the four-unit and six-unit concentration is certainly high enough and the similarity in market shares among leading suppliers sufficiently even to pass the criteria of Bain and other industrial economists for markets where sellers' collusion is feasible.[5] It may be noted

Chart 7.1: Country Concentration of Mineral Production: Cumulative
Share of Western World Output Accounted for by Four Largest
Producing Nations in 1974

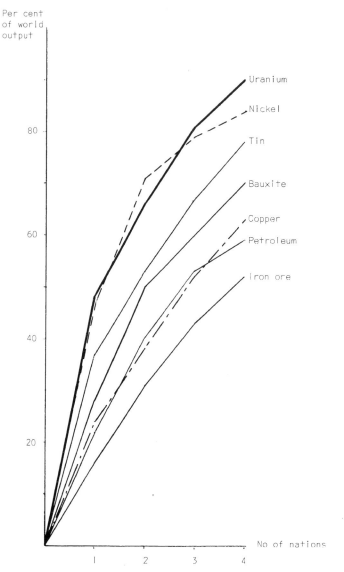

Sources: OECD, *Uranium Resources, Production and Demand* (OECD, Paris,
December 1975); BP, *Statistical Review of the Petroleum Industry* (London,
1978); *Mining Annual Review* (1976); *Metallgesellschaft*, Metal Statistics
(1977).

that though concentration appears to be considerably higher in the nickel and tin industries as compared to uranium, collusion in the former may be harder to achieve because of the great differences in the sizes of leading suppliers, and the consequent difficulty in creating a common ground for solidarity.

Table 7.1: Concentration in the World Uranium Market

Decision Unit	Locations	Estimated Production Capability at the End of 1977		
		Short Tons U_3O_8	Percentage of of Western World	Cumulative Percentage of Western World
RTZ	Namibia Canada Australia	8,000	19.0	19.0
Uranex/CEA	France Gabon Niger	6,000	14.0	33.0
Denison Mines	Canada	3,800	9.0	42.0
Kerr McGee	USA	3,800	9.0	51.0
Nufcor	South Africa	3,700	8.0	59.0
United Nuclear	USA	3,600	8.0	67.0
All others		14,100	33.0	100.0
World		43,000	100.0	100.0

Note: This is a highly approximate assessment compiled from a variety of sources including OECD, *Uranium Resources, Production and Demand* (OECD, Paris, December 1975), *Mining Annual Review* (1975-8); and RTZ, *Annual Reports*, 1975-9.

Transparency of transaction terms is of course greatest in cases where transactions are centred on an exchange like the LME which regularly publishes price and turnover figures. But the transparency condition is reasonably satisfied in the uranium market too. The buyers of uranium are relatively few in number. Most trade takes place under contracts covering such large volumes and such long time periods that they could not pass unnoticed by competing suppliers. Even if price details are often not published, deviations from joint agreements would be reasonably easy to detect for insiders in the small supplier group. Spot transactions, too, are closely monitored by bodies like NUEXCO and Nukem. This state of things would prevent persistent cheating and hence facilitate supplier collusion.

Table 7.2: Measures of Concentration in a Few Minerals Industries,
1977

| | Percentage of Western World Capacities[a] for | | | |
	Four Leading Corporate Decision Units	Six Leading Corporate Decision Units	Largest Corporate Decision Unit	Sixth-largest Corporate Decision Unit
Uranium	50	67	19	8
Alumina-refining	52	65	13	5
Copper-mining	38	49	11	6
Iron ore	37	52	12	7
Nickel-refining	72	83	37	5
Tin-smelting	69	83	30	6

Note: [a] For iron ore, the figures relate to 1977 production.

Sources: Uranium from Table 7.1. Other minerals from M. Radetzki and S. Zorn, *Mineral Processing in Developing Countries*, UNIDO/IOD, 328 (UNIDO, Vienna, 19 December 1979).

We turn secondly to a discussion of the circumstances which should prevail if supplier collusion with regard to price or quantity is to be profitable. At least three conditions have to be satisfied.[6]

The first one is that the colluding suppliers should account for a large enough market share. *Ceteris paribus*, the smaller this share, the more price elastic will be the demand for the collaborating producers' output, and the greater the proportion of sales lost as a result of an increase in the price for their supply. This condition is closely related to that of market concentration. The problems of reaching a collusive agreement will grow with the number of participating members. The condition that the colluding members cover a wide segment of the market, therefore, will be easiest to satisfy where concentration is high.

The second condition for profitable collusion is that overall demand should be inelastic to price, for if it is not, the users will turn to substitutes as the price is raised. The price elasticity of uranium demand is certainly low. As noted in Chapter 6, uranium constitutes a small fraction of the overall costs of nuclear electricity generation, while most of these costs are accounted for by capital investments. Since there are no substitutes for uranium in nuclear power generation, once the investment funds have been committed, uranium will be bought at any price within a very wide range, although reprocessing and variation in enrichment tails may afford the nuclear power producer some leeway in the amount of natural uranium needed. The relative inflexibility of uranium demand makes its price highly sensitive even to limited shifts in supply.

The third condition is a low price elasticity for supply from outside the colluding group. The application of this condition to the uranium market requires several comments. Considering first the shorter run, i.e. supply from existing installations, it may be noted that the elasticity of outsiders' supply is bound to be high so long as the industry operates below capacity. Non-participants to the market arrangement will be all too happy to expand their capacity utilisation in response to a price rise occasioned by cartel action, thereby reducing the benefits which accrue to the colluding participants. In contrast, the price elasticity of outsiders' supply will be very low, and the profits to colluding members from price increases or output restrictions much greater, when the industry operates at full capacity. The inventory situation is also important. Prospects for profitable collusion will be greater where stocks are small or controlled by the colluding group. Over a longer time period a higher price level is likely to induce the establishment of new mines outside the control of the collusive group. For this reason the price elasticity of outsiders' supply would typically be higher in the long run. In this context, however, the extended gestation period for uranium-mining investments must be taken into account. The cartel may find it profitable to raise prices and reap the temporary (three- to five-year) advantage of the low elasticity of outsiders' supply, despite the detrimental long-run effect caused by the creation of new production units. Barriers to entry may help in perpetuating the low elasticity of supply outside the cartel's control. Establishment of additional independent capacity will be much slower or will not take place at all if, as in the case of uranium, the colluding group has a tight control over a large proportion of reserves, or if it includes governments (Australia?) prepared to restrict the development of new mines.

Empirical observation of successful supply collusion suggests that substantive market action is usually preceded by an extended period of general collaboration among suppliers, on issues not directly related to profits, e.g. exchange of information, market studies or promotional campaigns. In this 'preparatory' period, the suppliers get acquainted, and develop a feeling of trust for each other. Without such trust, the group discipline required in collusive market behaviour would be difficult to bring about.[7] Observation also suggests that joint price-setting or supply-restricting action is often sparked off by externally induced upward shifts in demand. As will be apparent from the following, these observations, too, appear to be directly relevant to the evolution of the international uranium cartel.

A further point of considerable relevance to uranium needs to be brought up. Monopolistic pricing is not dependent on strictly formalised arrangements among suppliers. Thus, writing as early as 1933, E.H. Chamberlin concluded that when sellers are few and products standardised, a monopoly price can be established and maintained without formal collusion.[8] Chamberlin's results have been further elaborated upon by Bain and Scherer.[9] A tacit understanding enabling each supplier to behave in the collective interest of producers even in the absence of formal meetings and decisions will be particularly easy to maintain where the industry has had previous experience of formal cartel arrangements.

Chart 7.2 contains a graphic summary of the major elements explaining the consequences of successful market collusion such as occurred in the uranium market. D_1 represents the depressed level of demand for the potential cartel's output prior to 1973, while S is the marginal cost curve of the cartel producers. In the initial competitive conditions, price and the quantity supplied will be determined at p_1 and Q_1 by the intersection of D_1 and S. Note that in this situation operations render a loss, since average total costs, depicted by the ATC curve, are above the price received.

Assume now that events like those which occurred in uranium in 1973 and 1974 shift demand upwards from D_1 to D_2. Note that D_2 has been drawn steeper than D_1 to reflect the lesser potential of suppliers outside the cartel to expand output in periods of booming demand. With continued competitive supplier behaviour, the new price and quantity supplied would be p_2 and Q_2, determined by the intersection of D_2 and S. In the new situation, the industry is profitable, since the average total costs are lower than the price received.

Suppose, however, that the upward shift in demand converts the latent tendencies towards collusion in the supplier group into outright efforts to charge monopolistic prices. Maximisation of joint profits in the producer group will be attained by shifting output back to Q_3, given by the intersection of MR_2, the marginal revenue curve to suppliers after the upward shift in demand has occurred, and S, the industry's supply curve. With demand at D_2, price will rise to p_3 as a result of the curtailment in supply. It appears that with given conditions of demand, the disparity between the competitive price (p_2) and the monopolistic price (p_3) can be very substantial.

The chart is a convenient device for illustrating the difference

Chart 7.2: The Consequences of Successful Market Collusion

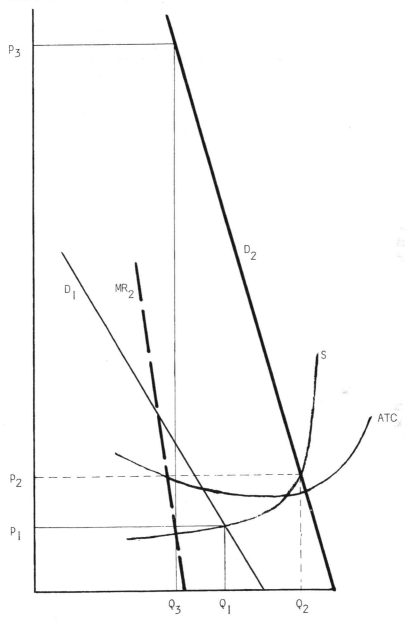

between competitive and monopolistic price. On the other hand, it should be clear that the neat and straight lines over-simplify the circumstances under which the uranium cartel operated. For instance, the chart does not reflect the continuous growth of supply and demand (as distinct from the sudden demand shift). Also, in reality, the monopolistic price level p_3 is unlikely to be attained both because of incomplete information about cost and demand conditions and because of unwillingness among members of the collusive group to submit themselves fully to collective decision-making.[10] However, given the limitation of established facts about the operations of the cartel, it is not possible to carry the formal analysis of the uranium case much beyond the stylised conditions and results depicted in Chart 7.2. To gain further insights under the circumstances, we turn in the following section to an informal analytical approach.

The Actual Operation of the Uranium Cartel

Although considerable amounts of material on the international uranium cartel have been uncovered, both in uranium litigations in the US involving Westinghouse[11] and in hearings in the US Congress investigating possible breaches of US antitrust legislation, part of the following discussion has to be based on reasonable deductions rather than on hard evidence. This should not be unexpected, given the actions of several governments and the cartel's participants to keep evidence about the uranium cartel secret, to avoid criminal prosecution and civil damages under the antitrust laws of various countries and the Treaty of Rome.

The existence of international cartel arrangements in the uranium market between 1972 and 1975 has been clearly admitted. It is evident that the Canadian government played an important role as an initiator of and member in these arrangements. Thus, in a letter of 15 August 1977 to the US Ambassador in Ottawa, the Canadian Acting Secretary of State writes:[12]

I have the honor to inform you that the policy of the Canadian Government was to support and participate in international uranium marketing arrangements from 1972 to 1975 to ensure the survival of the Canadian uranium industry which was being damaged by the restrictive uranium trade practices of the United States.

There is uncontested evidence of the existence of a formal cartel which organised meetings to set minimum prices and to divide the world market among participating member producers and nations during 1972 and 1973.[13] Given the strong government involvement in French and South African uranium production, the participation of producers from these two countries in the market arrangements implies at least a tacit support of their governments for the cartel endeavours.[14] Circumstantial evidence strongly suggests that several leading producers from the US took an informal part in the arrangements.[15]

The formal collaboration started while the market was weak and prices low. As indicated in Chapter 3 and in the above quotation of the Canadian Secretary of State, the depressed market conditions may have been the trigger which brought producers together in an effort to safeguard their own survival. Although competition among uranium suppliers virtually ceased early in 1973,[16] the bargaining power of the cartel appears to have remained modest until the end of that year. It took time to reach agreement and to work out firm institutional arrangements among the partners in the group. Also, since the cartel controlled less than the entire world output, any more drastic action to raise prices or cut supply might well have proved counter-productive so long as demand remained weak and outside producers were likely to underbid and supplement the supply of cartel members through expanded production or depletion of inventories. Thus, in June 1972, at its Johannesburg meeting, the cartel was striving for prices of $6.45 for 1975 deliveries,[17] i.e. not much higher than the 1972 market price. Though the appetite for, and the ability to force through, higher prices had been considerably improved by 1973, the minimum price agreed to by the cartel in October in that year, $7.70 for 1975 deliveries,[18] was still very modest compared to what was to come.

It is significant that after each of these meetings, the Canadian government directed its Atomic Energy Control Board to ensure that uranium from Canada was not sold at prices below the approximate level determined by the cartel.[19] Thus, in a direction attached to a letter of 17 August 1972 to the President of the Board, Donald Macdonald, the Canadian Minister of Energy, Mines and Resources, instructed, *inter alia*, that $6.45 should be the minimum price to be charged for Canadian uranium delivery commitments in 1975. In another letter, dated 30 October 1973, the same minimum price had been raised to $7.50. From a further letter of 6 March 1974, stipulating the minimum price for 1975 deliveries at $9.05, one may deduce that another cartel meeting had taken place in the early months of

1974. This government involvement clearly facilitated the adherence of Canada's uranium producers to the cartel decisions.

No public information is available on the further operation of the cartel. We do not even know for sure whether cartel members continued to convene in formal meetings to discuss prices and supply restrictions. Even though lack of evidence on this point may be decisive in a legal antitrust case, it should be clear from our discussion in the preceding section that the attainment of monopolistic pricing as illustrated by price p^3 in Chart 7.2 is not crucially dependent on formal cartel arrangements. This contention is, of course, not a judgement on the presence or absence of formal collaboration among producers in the uranium market after 1973. It is merely an indication that under existing market conditions and with the close relationships developed among cartel members and their governments in earlier years, the objective of monopolistic pricing should have been relatively easy to pursue, even if the formal cartal apparatus was scrapped in 1975.

Although there would have been substantial uranium price increases even in the complete absence of producer collaboration, it is our determined opinion that the uranium cartel, whether formal or informal, was highly successful in its actions to keep back supply, and to force prices to rise far higher than they would have risen if competition had prevailed among producers. This opinion is based on several inferences.

First, there is the interesting fact that output, which had grown continuously from its low mark in 1966, stagnated in 1973, and fell to lower levels in 1974 and 1975, despite the very encouraging price developments in these two years. As demand and prices increased, one would have expected producers acting independently to respond quickly by expanding production, to add to their depressed cash flows and to reduce average production costs.

As appears from Chart 7.3, it is not possible to explain the fall in output by constraints in production capacity. The data suggest that up to and including 1975, the industry operated far below existing capacity. Thus, in 1970, capacity utilisation, widely reported by the industry to be very low, can be assessed at 73 per cent. In 1973 and 1975 the figures were 73 and 75 per cent respectively. Mutual agreement among producers is necessary to explain why capacity utilisation remained low in 1974 and 1975, when prices were well on the way up.

It is, of course, possible to argue about the reliability of the OECD capacity figures which we use. For instance, in an interview early in 1980, representatives of RTZ suggested that at least in the US, official capacity figures for 1973 and 1975 included moth-balled installations which could not be brought to use without lengthy redevelopment. According to this view, capacity utilisation, as reported by OECD, could rise from 75 per cent in 1975 to 80 per cent in 1977, because only by the latter year was the formerly idle capacity ready to be operated.

This view, however, is contradicted by an authority intimately involved with the uranium industry. While strongly critical of the official US capability data, D.S. Robertson & Associates come to the conclusion that utilisation of available, ready-to-use capacity, more realistically measured by their own method, was 100 per cent in 1976, but that it was on average more than 12 per cent below that limit in the three preceding years.[20] This conclusion would appear to support our view that capacity utilisation was held back by producers rather than by physical constraints in the 1973-5 period.

Chart 7.3: Western World Uranium Production and Capacity in the 1970s

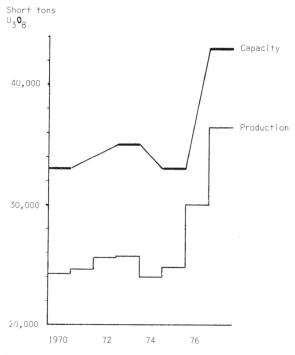

Short tons U_3O_8

Source: Chart 3.1 and Table 4.2.

Second, there is the widely publicised withdrawal of producers from the market in 1973 and 1974.[21] Inadequate reserves were frequently given as the reason for producers' unwillingness to enter into long-term contracts at this time. The argument sounds less than entirely convincing, given the state and development of reserves in the 1970s, as illustrated in Table 4.3. Producers' unwillingness to enter into commitments with fixed prices is fully rational if continued fast price increases were expected. Given the ample availability of reserves, the excess production capacity and the large and growing global uranium stocks, the producers' behaviour is difficult to understand unless one assumes that they were confident that ranks would not be broken and that their joint action guaranteed a continued rise in prices.

The producers' withdrawal from the market was very effective. Prices went up not only because supply as such was reduced, but also as a result of the ensuing nervousness among the unorganised consumers. Producers often required immediate acceptance of the sales offers they made, or else the offers were withdrawn. This added to the buyers' uncertainty. In aggregate, utilities and their governments had large stocks which would have lasted them a long time, if appropriately distributed. As it was, because of the general uncertainty about future uranium availability, those who were well supplied refused to lend to those who were not. The latter tended to panic in their attempts to obtain needed supplies, thereby driving prices to ever higher levels.

Third, the fact that the extreme rates at which prices were rising suddenly ceased after the quotation had reached the even figure of $40 (NUEXCO spot) in early 1976 is somewhat surprising, unless one assumes that $40 (nominal dollars) was the target price of the cartel. Once the target had been attained, producers could relax, re-enter the market and expand their capacity utilisation. This argument explains the sharp rise in production in 1976 and 1977, and is in line with D.S. Robertson's finding[22] that US producers' output reached peak capability in 1976.

One can only speculate about the circumstances under which a target price of $40 might have been established. The strength of the price rise took everybody, including the cartel members, by surprise. Some time during 1975 the colluding producers must have asked themselves when to discontinue the restrictive measures which were then in force. Forty dollars must have appeared as a round and exceedingly profitable figure, within reach under the circumstances that were evolving. Events moved so fast that it is improbable that the cartel gave any detailed consideration to the sustainability of a $40 price level. It is more likely that the target price was merely intended to be a signal that output restrictions and export quotas could be abolished. As it turned out, the speed of developments made the price overshoot the target.

There are further indications of close producer collaboration. To increase their grip over the market, producers agreed at an early stage in their collaboration to discriminate against uranium middlemen, e.g. reactor vendors, in their sales policies.[23] Later, while pursuing their primary price target, producers soon also turned to other objectives. Foremost among these were renegotiation of old low-priced long-term contracts, and the introduction of the market price concept combined with a high base price in new long-term contracts. The first move aimed at improving producers' receipts for their overall current supply. The latter had the purpose of conserving the high price level for a major proportion of the supply for a long time after an eventual weakening of the spot market had taken place. It is unlikely that these moves would have been taken by so many producers within such a short period of time, unless there had been collusion within the producers' group.

Conclusions

In this chapter we have tried to substantiate the claim that producer collaboration is an important element in understanding developments in the uranium market since 1973.

We started out by specifying the necessary preconditions for successful cartel action and found that these were indeed prevalent in uranium. We noted that producer collaboration is much easier to pursue when it is openly or tacitly supported by governments in producing countries, and found that such support did prevail in uranium. Experience from other fields suggests that collusive measures among suppliers aimed at raising prices and profits is often sparked off by exogenous upward shifts in demand such as that which occurred in uranium in 1973.

Our analysis of the functioning of the cartel until the end of 1973 was based on undisputed evidence of formal meetings to determine prices and divide markets. Analysis for the subsequent period had to be based on inferences and deductions derived from the functioning of the market. It is uncertain but economically inconsequential whether or not cartel collaboration was continued in 1974 and later through formal producers' meetings. Observation of market behaviour indicates that suppliers can maintain monopolistic pricing where market structure is conducive to profitable collusive behaviour, even in the absence of formal arrangements. This would be particularly true where, as in uranium, the suppliers have had ample opportunities to reach a basic understanding on how to pursue their collective interests during a preceding period of formal collaboration.

Our deductions suggest that the major tool of the cartel was to restrict supply until a target price of $40/lb U_3O_8 (nominal dollars) had been firmly accepted by the buyers. We believe that prices would never have reached anywhere near that level if uranium producers had been competing with each other. On the other hand, we feel unable to provide a quantitative measure of the impact of the cartel in isolation. It is clear that producers' action alone would not have sufficed to accomplish what actually happened. The sharp jump in uranium demand in 1973/4, caused primarily by US enrichment policies at the time and the emergence of appropriable quasi rents following the oil price rise, were necessary prerequisites which breathed life into the cartel and boosted the producers' bargaining strength.

Further factors, both external and internal to the uranium market, emerging from 1974 onwards have been instrumental in supporting prices, and in permitting producers to maintain a grip on the market, despite the depressive influences of falling forecasts of nuclear capacity expansion and decreased requirements of uranium due to adjustments and cancellations in enrichment contracts. In the next chapter we analyse the nature and consequences of these market-boosting factors.

On balance, it seems that the producers' market power was weakened after 1977. While the target price of $40 (nominal) has been maintained, producers have been unable to prevent a substantial erosion of real price levels in 1978 and 1979.

We end this chapter by drawing attention to the similarities between the uranium cartel and OPEC. Like OPEC, the uranium cartel had strong government participation. Like OPEC, the uranium cartel was initially created for the defensive reason of preventing low prices from falling even further.[24] Like OPEC, the uranium cartel was unable to influence prices in any drastic way until factors external to the cartel had created the right preconditions for market and price control. Finally, as in the case of OPEC, producers, following an extended period of organised collaboration, were prepared to seize the opportunity, once the right preconditions presented themselves. But while continued high levels of petroleum demand, coupled with reduced suppliers from major producing countries, permitted OPEC to keep the real price of oil between 1974 and 1978 at a relatively stable level, and to increase it considerably in 1979, reductions in current and forecast uranium demand, together with the sharp upward jump in supply after 1975, have reduced the scope of market and price control exercised by uranium producers.

Notes

1. J.S. Bain, *Industrial Organization* (John Wiley, New York, 1968), p. 119.

2. R.E. Worrol, 'Uranium Mining and Marketing in South Africa' in *Natural Uranium Supply* (International Symposium of Deutsches Atomforum, Mainz, November 1974).

3. *Annales des Mines* (July-August 1976), pp. 130-41.

4. Ibid.

5. Bain, *Industrial Organization*. See also F.M. Scherer, *Industrial Market Structure and Economic Performance* (Rand McNally, Chicago, 1970), and J.F. Pickering, *Industrial Structure and Market Conduct* (Martin Robertson, London, 1974).

6. For a fuller discussion see M. Radetzki, 'The Potential for Monopolistic Commodity Pricing by Developing Countries' in G.K. Helleiner (ed.), *A World Divided* (Cambridge University Press, Cambridge, 1975), and Carl Van Duyne, 'Commodity Cartels and the Theory of Derived Demand', *Kyklos*, vol. 28 (1975).

7. For a further discussion see Scherer, *Industrial Market Structure*, Chapter 6.

8. E.H. Chamberlin, *The Theory of Monopolistic Competition* (Harvard University Press, Cambridge, Mass., 1933), Chapter 3.

9. Bain, *Industrial Organization*, Chapter 9, and Scherer, *Industrial Market Structure*, Chapter 5.

10. For an interesting theoretical and empirical discussion of the internal dynamics of commodity cartel operations, see Paul Eckbo, *The Future of World Oil* (Ballinger, Cambridge, Mass., 1976).

11. *Pittsburgh Post Gazette*, 6 January 1977.

12. See Government of Canada, 'Policy Statements and Press Releases on Uranium Marketing' (Ottawa, undated).

13. *Financial Times*, 4 September 1976. See also several following references to documents which describe individual cartel meetings.

14. In a uranium trial between Homestake Mining and Westinghouse, concluded in December 1979, a US Federal Court noted that 'Uranex [the French uranium marketing agency] is partly owned and largely controlled by the French government.' In 1974 and 1975, notes the Court, 'Uranex began to receive pressure from the French government concerning the huge disparity between [existing] contract prices and the going market price, and Uranex was forced to renegotiate all its contracts with buyers.' See *Nuclear Fuel* (21 January 1980).

15. Gulf Oil Corporation of the US was successfully prosecuted by the US Justice Department for taking part in the cartel. See *An Analysis of Competitive Structure in the Uranium Supply Industry* (Federal Trade Commission, Washington, DC, August 1979), p. 41.

16. Euratom Versorgungsagentur, *Annual Report 1973*.

17. Minutes of the meeting, presented as evidence with reference no. H10 in legal case T192-194/75 at the District Court of Stockholm, Dept. 7.

18. Minutes of cartel meeting held in London, 8-9 October 1973, presented as evidence with reference no. H17 in legal case T192-194/75 at the District Court of Stockholm, Dept. 7.

19. See Government of Canada, 'Policy Statements'.

20. D.S. Robertson & Associates Ltd, 'Uranium Price Movement and the

Reasons Therefor' (August 1977), p. 18.

21. See, for instance, *NUEXCO Reports*, no. 62 (20 September 1973), no. 63
22 October 1973), no. 67 (19 February 1974); Statement of J.E. Gilleland of
TVA at hearings before the Joint Committee on Atomic Energy, US Congress,
September 17 and 18 1974, and Euratom Versorgungsagentur, *Annual Reports*,
1973 and 1974.

22. Robertson & Associates, 'Uranium Price Movement'.

23. Minutes of cartel meeting held in Johannesburg in June 1972, presented
as evidence with reference no. H10 in legal case T192-194/75 at the District Court
of Stockholm, Dept. 7.

24. *Financial Times*, 4 September 1976.

8 OTHER FACTORS RESTRAINING SUPPLY IN THE 1973-1980 PERIOD

Introduction

The preceding chapters have analysed the impact on the uranium market arising from the policies with regard to reprocessing and enrichment, from the oil crisis and from the international uranium cartel. Several miscellaneous factors with a strong influence on uranium market developments remain to be treated.

In the following we discuss the Australian policies pertaining to the development of that country's uranium industry, the export policies of the government of Canada, the French withdrawal as supplier of uranium following political changes in Niger, the Westinghouse announcement of its inability to honour its contract commitments, and the actions of Australia, Canada and the US, motivated by their desires to avoid nuclear proliferation. All these factors have had the effect of strengthening the sellers' market position and helping them to push prices upwards or keep the price levels high through real or threatened reductions in international uranium supply.

The bundling together of several factors into the present chapter is not a reflection of their lesser importance, as compared to those treated in preceding chapters; the reason is rather that the analysis of each requires a relatively limited amount of space. Furthermore, it should be clear that the factors to be discussed in the following occurred during different periods of time, sometimes simultaneously with or even prior to the events explored in earlier pages.

Australian Production Delays

Very large and rich uranium deposits were discovered in Australia in the early 1970s. By the end of 1972, most of these bodies were ready for development. The long-run world supply forecasts of these years attributed a very important role to future production in Australia. Then, late in 1972, there was a change in government, and with it a drastic and prolonged change in uranium development prospects. Writing in 1977, Grey notes[1] that 'the state of paralysis from which the industry has suffered since 1972 continues.'

The consecutive delays in the development of Australia's uranium production have been primarily caused by the government's desire to keep the industry under national control, the trade union and aboriginal opposition, environmental worries and, more recently, export restrictions related to Australia's non-proliferation policies.[2] Table 8.1 summarises the ensuing delays in the country's production plans. As late as mid-1973, large-scale supply was expected by 1978. Two years later, the plans had been postponed to the early 1980s, and in 1978 an optimistic production forecast put the 1982 output level at no more than 4,600 short tons.[3]

The impact of the Australian delays on the uranium market has been substantial. In 1973, an aggregate production in the five-year period 1978-82 of some 45,000 short tons was expected.[4] In 1978, the expected production volume for the same years had been reduced by 37,000 short tons to only 8,000 short tons.[5] Demand for future delivery had to be redirected from Australia, thereby causing increased demand pressures on other supply sources. This, in turn, must have improved the suppliers' bargaining position regarding future prices, and, by implication, facilitated their efforts to raise spot prices too.

Table 8.1: Australian Reserves and Production Plans

Period of Observation	Reserves Cost Category below Current US $	Short Tons U_3O_8	Planned Production Capacity (short tons U_3O_8)				
			1973	1975	1978	1980	1982
1970	10	22,000	1,500				
1973	10	92,000		1,000	6,000		12,000
1975	15	243,000			1,000	4,200	6,500
1977	30	290,000			700	700	6,000

Sources: OECD, *Uranium Resources, Production and Demand* (OECD, Paris, September 1970); ibid. (August 1973); ibid. (December 1975); ibid (December 1977).

The delaying measures of the Australian government were of course very much in line with the policies of the international cartel, and the question of whether the government took an active part in the cartel activity has been raised. Though Australian mining companies with uranium reserves did participate in the early cartel meetings, it is somewhat difficult to see an economic justification for the government of

Australia to support the market arrangements by deferring the development of the country's deposits. Had the government been a cartel member, a rational policy on its part would have been to secure a market share for Australian uranium, and to benefit from the high prices by a speedy development of production capacity to fill the country's quota.

Australia is now expected to become a large-scale uranium producer by about 1985. At that time, its share of world output is assessed at 17-18 per cent, second only to that of the US.[6] Clearly the substantial addition to world supply implicit in Australia's entry on a large scale will complicate the producers' efforts to control the world market, and will add to the depressive price tendencies felt in the late 1970s. The Australian Atomic Energy Commission, for instance, fears that there may be an over-supply of uranium to 1985 if the large potential for new production is realised.[7]

Canada: Export Restrictions and Price Manipulation

The Canadian government has been actively involved in various ways in the country's uranium-mining ever since this industry was established. Apart from (or maybe as part of) its involvement in the uranium cartel as described in Chapter 7, the Canadian administration has pursued two policies with a very significant impact on the uranium market in the post-1973 period.

In September of 1974, the Minister of Energy, Mines and Resources introduced a policy package which purportedly aimed at assuring uranium supplies to the national nuclear development programme. One of the requirements of this package was that the mining firms set aside uranium reserves adequate to cover thirty years' forward requirements of uranium in all existing and planned nuclear reactors in Canada. Another was to refuse approval for firm export contracts beyond ten years into the future.[8]

Analyses presented by the Canadian Department of Energy, Mines and Resources as of January 1975 indicated that out of the then existing reserves, measured by the government for this purpose at 400,000 short tons, more than 90,000 short tons would have to be set aside immediately to assure future Canadian requirements. The analyses also showed that on the basis of existing reserves there would be no scope for further exports from Canada from 1984 onwards, since at that time all these reserves would have to be set aside for future domestic consumption. To assure future domestic needs by the year 2000 in the way stipulated by the policy, Canadian miners would have to create and set aside reserves amounting to more than 900,000 short tons.[9]

When seen as a measure to assure Canada's own uranium requirements, the policy package must be regarded as unrealistic and exaggerated. The country's size and proven history of mineral discovery make the likelihood of uranium depletion within any reasonably foreseeable future a very remote possibility. The requirement to set aside expensively defined reserves to the extent envisaged appears uneconomic and extravagant, though not without precedent in government policy.[10]

Given the experience and sophistication of the Canadian administration in the field of minerals, an alternative rationale to the policy decisions presents itself. This rationale would be to give a further upward boost to the rising uranium price trend by *suggesting* that Canadian export supplies could not be assured for the longer term and that these supplies might soon dwindle. We underline the word 'suggesting', for although the policy did arouse strong apprehensions among consumers,[11] thereby making them more willing to accept the increasing prices demanded by producers, the period since 1974 has seen very large and rich additions to Canadian reserves,[12] pushing the time when the policy implementation would restrict exports into the more distant future. It is probable that the requirements for the Canadian producers to keep reserves aside for domestic needs, as envisaged by the policy, will be abandoned in coming years.[13]

The contention that the rationale for the Canadian export restriction policy package of 1974 was simply to support the price rise is in line with the second policy pursued by the government of Canada in more recent years. The clear-cut aim of the latter was to boost uranium prices. Thus, in March 1977, i.e. two years after Canada had purportedly discontinued its involvement in international uranium market arrangements,[14] the Minister of Energy, Mines and Resources declared that the government expected all future uranium contracts to contain an escalating floor price, and to provide for annual price renegotiations, to assure that Canadian price levels would never fall below the 'world market price'. He also encouraged producers to renegotiate their older low-priced contracts.[15] The Canadian administration even went beyond the issuance of general policy directives, and intervened to raise prices in specific contracts. Thus, it is reported that late in 1977 the Canadian administration forced through a price increase in two long-term contracts to above the price level which had been initially agreed upon between the Canadian producers and their Italian and Swedish customers, and approved by the Canadian government.[16]

The Canadian policies described here supported the upward price rise from 1974 onwards by reducing expected export supply, and helped in maintaining the high price levels from 1977 onwards, by the government's administrative interventions in price formation. Hence, these policies were fully in line with the objectives of the international uranium cartel. In fact, one might well surmise that in spite of official denials, the policies formed part of the Canadian government's continued active role in the formal or informal international collaboration of uranium producers.

The French Withdrawal from the International Market

An event of smaller quantitative importance but highly significant in timing was the withdrawal of France early in 1974 as a supplier to the international market. This was reported to have been caused by failing supplies from Niger and Gabon during most of the year,[17] following a *coup d'état* in Niger in March 1974, and subsequent demands by the governments of the two countries that prices for their deliveries be increased in line with world market developments.[18] Though the shortfall caused by these events cannot have exceeded Niger's and Gabon's 1974 production (about 2,000 short tons), the withdrawal of France from the market soon after OPEC's price increase came at a very sensitive time, and can be taken to have added fuel to the rise in uranium prices in that year.

The Westinghouse Delivery Failure

In September 1975, Westinghouse declared its inability to deliver uranium to a number of utilities in the US and Sweden. The total quantities involved have been estimated at about 32,000 short tons.[19] This includes some quantities for which the existence of delivery contracts has been disputed. Deliveries are spread over a prolonged period of time, i.e. from 1976 until 1993. Delivery commitments for the 1980s and beyond are especially difficult to calculate, since they are conditional upon the utilities' future needs, as these may develop over time.

To assess the quantitative impact on the market of the reduction in anticipated supply due to Westinghouse's failure, we measure the emergent shortfall over a five-year period, just as we did earlier when discus-

sing postponed recycling, fixed enrichment contracts and delayed production in Australia. It appears that the Westinghouse shortfall is greatest in the period between 1977 and 1981, when it amounts to 16,600
short tons.[20]

Like the many other factors discussed in this section, Westinghouse's
inability to deliver has added to the upward push in uranium prices, by
redirecting demand to other sources of supply. But further considerations have to be brought out for a proper evaluation of the strength and
nature of this impact.

Most important is the fact that since only a minor share of Westinghouse's supply commitments was to come from its own uranium production endeavours, a major proportion of the uranium to be supplied
had to be purchased by Westinghouse in the market some time before
delivery. It follows that if Westinghouse had fulfilled its commitments,
anticipated supply to final customers would have remained unchanged,
but the market would have received an upward push as the company
covered its uranium shortfall. With its declaration, Westinghouse reduced anticipated supply, with a consequent strengthening of the market.
But by relieving the company of the need to buy uranium, the declaration simultaneously resulted in a reduction in anticipated market
demand, with a consequent weakening of the market. Assuming full
information and rational market behaviour, the above suggests that
Westinghouse's uranium endeavours, in so far as they related to quantities that the company had to purchase in the market, had the double
effect of both strengthening and weakening the market, with the two
forces offsetting each other.

Admittedly, market agents do not always behave rationally. This
would be particularly true in the tense uranium market situation,
characterised by the galloping price increases which prevailed both prior
to and after September 1975. Besides, although Westinghouse's short
uranium position is reported to have been widely known in the nuclear
industry since early 1974,[21] one can take it that information among
market agents about the extent and details of the company's uncovered
commitments was less than complete. One can therefore conjecture
that Westinghouse's uranium market activities slowed down the price
rise while the company went short, and increased its pace in the months
after September 1975, but that the final price level which emerged in
1976 was not strongly affected by the Westinghouse endeavours. On
top of this change in the price rhythm due to incomplete market information, there must also have been a purely psychological reaction to
the September declaration, which temporarily strengthened the bull

market in uranium.

An institutional consideration is also worth mentioning in this context. It is clear that its inability to fulfil its commitments virtually wiped out Westinghouse as a supplier of uranium. A consequence of this was an increase in the utilities' direct dependence on the uranium-mining companies. This in turn probably improved the cartel's ability to control the market. One can therefore argue that Westinghouse's failure, leading to its elimination as a uranium supplier, may have strengthened the price-boosting efforts of the cartel.

The Policies of Nuclear Non-proliferation

The nuclear non-proliferation issue has played an important role in the uranium market from 1977 onwards. Canada, the US and Australia have tried to force the uranium-importing countries to agree to a variety of rules with regard to nuclear practice as a precondition for continued uranium supplies.

The countries in Western Europe and Japan, the main importers of nuclear fuels, have been unwilling to comply with the national nuclear non-proliferation policies of the exporting nations. These policies are regarded as an unwarranted interference in the internal affairs of the importers. They are seen in many instances to be inconsistent with national legislation of the importing countries. For example, while the US non-proliferation policy appears to ban new nuclear reprocessing facilities, permission for the construction of nuclear plants in Germany is conditional upon the availability of reprocessing.[22] Given the continued heavy reliance of Germany on US nuclear fuel services, the German plans to establish an integrated reprocessing and waste disposal plant may be thwarted by a US threat to interrupt nuclear supplies.[23] In many cases, the policies of the exporting countries have also been inconsistent with each other. This has given rise to so called 'double-labelling' problems — situations in which fuel importers must live up to export conditions imposed by both uranium producers and uranium enrichers. This problem was experienced by Japan when Canadian uranium for Japanese use was enriched in the US, and both the US and Canada wanted to impose their national non-proliferation conditions on Japan's use of that uranium.[24] More recently, the 'double-labelling' problems have been partly solved by a new US-Canadian co-operation agreement which leaves it to the US to handle these matters in consultation with Canada.[25]

The importers' unwillingness to comply resulted in a Canadian uranium export embargo to the European Community and Japan during the whole of 1977.[26] Exports were resumed early in 1978, on the basis of interim agreements which leave the long-run supply uncertainties unresolved. Similarly, the US non-proliferation policy package introduced in April 1977 led to a temporary export embargo early in 1978, not on uranium, but on various nuclear services, to the European Community, in response to the latter's unwillingness to comply.[27] Though a compromise has been reached, the issues between the US and the European Community have not been permanently settled. Australia, too, is in the process of negotiating non-proliferation agreements with a variety of countries.[28] Such agreements, it appears, will be a precondition for future Australian uranium deliveries. While reasonable accommodation appears within reach between Australia, Canada and the US on the one hand and the uranium fuel-importing countries within the OECD area on the other, supply restrictions due to non-proliferation appear to be more severe for deliveries to Third World importers. There is a tendency among these, therefore, to concentrate their uranium demand on South Africa, Niger and Gabon, whose export policies in this field are much more relaxed.[29]

The non-proliferation issue has had a direct impact on uranium supply. The Canadian export embargo led to a delivery shortfall of 3,300 short tons to the European Community[30] and of further quantities to other European countries and to Japan. More important than the physical reduction has been the psychological impact on uranium importers of this new element of supply insecurity. Unwilling to surrender to what is increasingly being regarded as a joint Australian-Canadian-US government arrangement.[31] the importing countries have increased the level of their desired uranium inventories,[32] and are diversifying their sources of supply by offering to pay higher prices for deliveries from countries other than the three.[33] Knowledgeable market observers contend that for this very reason the 1979 prices received by suppliers from Niger were about 5 per cent above the NUEXCO exchange value.[34] This would correspond to a premium of more than $2 per lb U_3O_8.

In summary then, the non-proliferation policies have provided support to uranium prices from 1977 onwards, by temporarily reducing physical supply, by increasing the level of desired stockholding and by making importers prepared to pay more to avoid the risks of supply disruption on non-proliferation grounds.

Notes

1. A.J. Grey, 'Current Australian Uranium Position', *Uranium Supply and Demand* (The Uranium Institute, London, 1977).

2. For a discussion of these issues, see W.B. Rotsay, 'Conditions Applying to Australian Uranium Exports', *Atomic Energy* (April 1976), and the articles on Australia in the Uranium Institute's *Uranium Supply and Demand* by A.J. Grey (1976 and 1977 issues), and by B. Lloyd and B. Frisk (1978 issue). The impact of the non-proliferation policies will be dealt with in a later section.

3. B. Lloyd, 'An Australian View of the Uranium Market', *Uranium Supply and Demand* (The Uranium Institute, London, 1978). The Uranium Institute's forecast of January 1979 puts Australian production in 1982 at between 1,800 and 5,000 short tons U_3O_8. See *The Balance of Supply and Demand 1978-1990* (The Uranium Institute, London, 1979), p. 39.

4. OECD, *Uranium Resources, Production and Demand* (OECD, Paris, August 1973). Production is assumed to grow in a linear function from 6,000 short tons in 1978 to 12,000 short tons in 1982.

5. Lloyd, 'An Australian View'.

6. *The Balance of Supply and Demand 1978-1990*.

7. Australian Atomic Energy Commission, *Annual Report* for year ended 30 June 1979.

8. See R.M. Williams, 'Uranium Supply to 2000, Canada and the World', paper presented at a meeting of the Geological Association of Canada, Edmonton, Alberta, May 1976.

9. Tables and charts of the Department of Energy, Mines and Resources, contained in Williams, 'Uranium Supply to 2000'.

10. In 1939, the Australian government imposed an embargo on iron-ore exports, so as to prevent the early exhaustion of national reserves, then assessed at about 450 million short tons. In 1960, export permission was restricted to one-half of newly found iron ore. (See M.A. Adelman, 'Economics of Exploration for Petroleum and Other Minerals', *Gevexploration*, no. 8 (1970). By 1978, iron-ore reserves in Australia had risen to some 20,000 million short tons (US Bureau of Mines), and exports in that year alone were 83 million short tons (*Mining Annual Review* (1979)), making the country the world's leading exporter.

11. Euratom Versorgungsagentur, *Annual Reports* (1974 and 1975). *Nukem Market Review* (September 1974).

12. See, for instance, R.M. Williams, 'Uranium', *Canadian Mining Journal* (February 1978).

13. The Canadian uranium industry is expressing increasing opposition to the public regulation of uranium exports. See *Nuclear Engineering International* (January 1980), p. 30.

14. Government of Canada, 'Policy Statements and Press Releases on Uranium Marketing' (Ottawa, undated).

15. Williams, 'Uranium'.

16. *Nukem Market Review* (July and August 1977, April 1978).

17. Euratom Versorgungsagentur, *Annual Report 1974*.

18. F. Oboussier, 'Supplying the Six', *Nuclear Active* (July 1975).

19. *Significant Events in the Uranium Market 1969-1976* (NUEXCO, October 1976).

20. Private communication with Westinghouse Corporation.

21. *Nucleonics Week*, 17 July 1975.

22. A. von Kienlin, 'Commercial Effects of Current Non-proliferation Policies', *Uranium Supply and Demand* (1978).

23. *SIPRI Yearbook 1979* (Taylor & Francis Ltd, London, 1979), p. 323.

24. *Nukem Market Review* (July 1977, January 1978).

25. *SIPRI Yearbook 1979.*

26. *NUEXCO Market Report* (January 1978).

27. Von Kienlin, 'Commercial Effects of Current Non-proliferation Policies'.

28. *Nukem Market Review* (August 1978).

29. *SIPRI Yearbook 1979.*

30. Euraton Versorgungsagentur, *Annual Report 1977.*

31. *Nukem Market Review* (May 1977), and von Kienlin, 'Commercial Effects of Current Non-proliferation Policies'.

32. *Nukem Market Review* (November 1978), and *Jahrbuch der Atomwirtschaft,* (1979), p. A40.

33. Von Kienlin 'Commercial Effects of Current Non-proliferation Policies'.

34. *NUEXCO Market Report* (January 1979).

9 CONCLUSIONS AND PROSPECTS

Conclusions

Uranium prices exploded in the mid-1970s. From 1973 to 1976 the quotation rose more than sixfold in nominal terms, and almost fivefold when measured in constant dollars. The market changes were general. Prices rose more or less in parallel in all major producing and consuming countries. The price changes affected current quotations for spot as well as for future deliveries. Producers pressured consumers into widespread renegotiations of old long-term contracts, whereby the low prices for future delivery, agreed upon in the past, were increased.

We compared the uranium price developments with the price histories of a wide group of major primary commodities in international trade. Going as far back as 1920, and excepting petroleum in the 1970s, we have been unable to document any price boom experienced by an individual commodity that could match that of uranium in terms of magnitude (both nominal and real), speed and generality of the price rise, as well as the perseverance of the high price level which followed the rise. This points to the uniqueness of the uranium market's price performance in the mid-1970s.

The uranium industry was originally developed in response to military demand. In the period between 1968 and 1973, the market for uranium was heavily depressed as a result of speedy curtailment of military procurements before matching commercial demand had developed. Capacity utilisation remained low during this period. Production was far above current requirements, and inventories were rising. The quantity of low-cost mineral reserves was expanding and remained comfortably adequate for future production planning. Forecasts of a very fast growth in future commercial uranium needs constituted a strong encouragement to producers. The prices which prevailed (between 1971 and 1973, the price in constant 1975/6 dollars varied between $8 and $10) were too low to enable producers to cover their full costs including a satisfactory return on invested capital.

The plans for expansion of nuclear capacity reached a peak in 1970. Each consecutive forecast made since then has implied a scaling down in the expected growth of nuclear installations, as compared with earlier forecasts. Even with the reduced forecasts of 1973, it was clear that uranium production and capacity would have to start expanding at fast rates from the latter 1970s, to provide for the growing nuclear needs. A number of studies, based on 1970-3 data, were undertaken to establish the incentive price level thought to be necessary to induce the uranium-mining industry to grow in line with the increasing requirements. These studies showed that the incentive price figures needed to assure a net-of-tax return of 12-15 per cent to marginal new uranium projects were bounded on the upper side by a price level of $19 (1975/6 dollars).

Neither the market nor knowledgeable industry opinion in the early 1970s expected uranium prices during the 1970s and early 1980s to rise as high as this incentive price level, however. Up to mid-1973 it was possible to sign long-term contracts for deliveries in 1980 at a price of no more than $10 (nominal dollars). And the consensus of opinion in authoritative forecasts made between 1971 and 1973 was that uranium prices in the early 1980s would be in a band of $12-14 (1975/6 dollars). These price forecasts were indeed far away from the levels of $40 and above actually attained in 1976.

The strength and perseverance of the uranium price rise was surprising, not least because falling demand forecasts, excess plant capacity, production above current needs, rising inventories and ample and fast-growing reserves continued to characterise the industry right through the period during which prices exploded. It is necessary to venture beyond these 'normal' market circumstances in order to catch the forces which shaped the uranium market developments from the end of 1973 onwards. The factors which in our opinion were mainly responsible have been discussed in detail in Chapters 5-8. They are presented again in summary form in Table 9.1.

Given that most of the factors listed in the table arose independently of one another, it is not surprising that the uranium market forecasters failed even late in 1973 to predict what was to come. Perceptive observers might have incorporated the possible impact of one or two of the items in their predictions. But to have predicted correctly the more or less simultaneous occurrence of all the ten factors and their combined consequences for the uranium market appears as a virtual impossibility.

In the table, we try to distinguish between those factors which had a real and immediate influence on the spot market and the ones whose impact was in the main of a psychological nature. The consequences of the latter factors (including partly or entirely items 2, 4, 7, 9 and 10) should not be underestimated, for they caused important alterations in expectations, which directly affected the futures market, thereby influencing demand and price in the spot market too. We also distinguish between three main types of market impact caused by the factors dis-

Table 9.1: Major Factors behind the Uranium Price Boom after 1973

No.	Factor	Period of Major Impact on Market	Real versus Psychological Impact	Nature of Impact	Quantitative Assessment of Impact, where Possible
1.	Delays in recycling and plutonium use	1974-7	Real	Increase in demand	Aggregate demand in five-year period 1978-82 increased by some 61,000 short tons
2.	Intentions to increase enrichment tails	1973-7	Psychological	Increase in demand	Aggregate potential demand in five-year period 1978-82 increased by about 20,000 short tons
3.	Fixed enrichment contracts	1973-8	Real	Increase in demand	Aggregate excess demand due to fixed enrichment contracts: In the US (assessed in 1976) for five-year period 1977-81, 46,000 short tons In Europe (assessed in 1977) for five-year period 1979-83, 43,000 short tons
4.	The oil crisis	1973-5	Mainly psychological Real	Increase in demand for higher inventories Directly on price through higher profits due to higher electricity prices	
5.	The uranium cartel	1973-7	Real and psychological	Restricted supply. Increase in demand for higher inventories	
6.	Delays in Australian production plans	1973-8	Real	Reduced supply	Forecast aggregate supply in five-year period 1978-82 reduced between 1973 and 1978 by 37,000 short tons
7.	Canadian government interventions re export volume and price	1974-8	Mainly psychological Real	Threatened reduction in supply Directly on price through administrative interventions	
8.	French withdrawal as supplier due to political changes in Africa	1974	Real	Reduced supply	Supply reduction in 1974 not exceeding 2,000 short tons
9.	Westinghouse's contract failure	1975-7	Real and psychological	Reduced supply	Aggregate supply commitments failure in five-year period 1977-81, 16,600 short tons
10.	Major exporters' non-proliferation policies	1977-9	Mainly psychological Real	Threatened reduction in supply. Increased demand for inventory build-up. Direct effect on prices for deliveries from other countries. Reduced supply due to Canadian embargo	Shortfall in Canadian deliveries in 1977 to the European Community, 3,300 short tons

cussed, viz. (a) increase in demand; (b) reduction in supply, each leading to an upward push on price; and (c) a direct effect on price independent of prior changes in demand and supply.

Another useful distinction among the factors may be between the ones that were internal to the uranium market and those that were not. To an economist, internal market forces would ordinarily include the supply, demand and price changes caused directly by the decisions of the agents in the market, in pursuit of their utility or profit maximisation goals, be they primary producers, intermediaries or final consumers. With this definition, only items 5 and 9, the uranium cartel's actions and the Westinghouse failure, clearly fall into the internal market factor category. All the remaining ones resulted in the main from actions and initiatives external to the uranium market. On closer scrutiny it appears that the eight external factors emerged as a consequence of political decisions and interventions by governments of energy-supplying countries. These affected the uranium market indirectly, through spillover effects in the case of item 4, the oil crisis, and directly in all other cases. Thus, as detailed in earlier chapters, items 1-3 were the result of administrative decisions taken primarily on other than economic grounds by government agencies in the US. Item 6, delay in Australian production start-up, was caused by a change in the Australian government's attitude towards uranium resource development. Item 8 is clear-cut: political changes in Africa temporarily deprived France of a substantial part of its uranium supply. Items 7 and 10, finally, are the consequence of national goals introduced and pursued by the governments of three supplying countries, causing serious disturbance to the buyers, and to a lesser extent also to the sellers in the international uranium market.

In seven cases we have succeeded in obtaining partial or total quantitative measures of the impact on supply or demand of the factors studied. In five of these (items 1, 2, 3, 6 and 9), where demand or supply are known to be affected over extended periods of time, and where the overall total may be hard to assess in an unambiguous way, we present the quantitative impacts in terms of supply reductions or demand additions over a specific five-year period, varying between 1977-81 and 1979-83, depending on the availability of data needed for the assessment. In this way we derive a rough measure whereby the importance of these factors can be compared.

The combined effect of the items summarised in Table 9.1 was clearly exceptional. Though aggregation is hazardous, we may obtain a first feel of the orders of magnitude by adding together those five factors whose five-year impact is quantified in the table. The overall five-year shift in market balance caused by these factors adds up to more than 223,000 short tons, which works out at an annual impact of almost 45,000 short tons, a quantity far above overall reactor requirements in the late 1970s (in 1978, reactor requirements amounted to 34,000 short tons). To this must be added the impact of the five further factors listed in the table, which was certainly important but which we have been unable to measure in quantitative terms.

The price-boosting impact of the ten factors was counter-balanced only in part by the reduced future uranium requirements due to lower forecasts for nuclear power expansion. As pointed out in Chapter 4, the 1978-82 nuclear capacity forecast made by the OECD in 1977 was 46 per cent lower than the corresponding forecast made in 1973. We assessed the consequent reduction in average annual uranium requirements in the 1978-82 period at 35,000 tons.

Even though we are unable to provide a clear-cut quantitative measure of the overall difference between the price-boosting impact of the ten factors listed in Table 9.1 and the price-depressing consequence of the reduced nuclear expansion plans, it is clear even from the partial quantitative assessment presented here that the former was far larger than the latter, and that the net impact was strong enough to explain the actual price performance in the uranium market.

Applications to Other Commodity Markets

As already noted, the price performance in uranium puts this commodity market together with that of petroleum in a special category. The conditions which led to the uranium price explosion were exceptional in several respects. Even so, there are at least two lessons to be drawn from our uranium analysis above which appear to be of general relevance for international commodity markets.

The first one is that market behaviour is difficult to judge and predict with the help of standard economic criteria when governments are heavily involved in regulating or operating commodity markets. Neglect of this fact was probably the major fallacy in the erroneous price forecasts made in the early 1970s about developments in the uranium market. The forecasts assumed that economic criteria, such as comparative costs, profitability and consumer utility, would govern these developments. In fact, events have been shaped to a considerable extent by political decisions which may have been, but more likely were not, guided by economic criteria. This is probably true of the introduction of highly complex regulatory procedures for nuclear power, which led to a sharp slow-down in nuclear expansion. It is certainly true of the governmental decisions in the US, Canada and Australia pertaining to recycling, enrichment, development of new uranium mines, and to the variously motivated export restrictions.

The generality of the problem becomes apparent when it is realised how widely governments have penetrated international commodity markets. In a recent study, R.F. Mikesell notes that[1]

> The ownership structure of several important non-fuel minerals industries in the developing countries has changed substantially over the past 30 years and particularly during the past decade. Specifically, ownership has shifted from foreign (usually private) mining companies to national (usually government owned) enterprises.

The world copper industry may provide a concrete example of current trends. In the 1960s, copper production in which government held any sort of interest represented no more than 2.5 per cent of non-socialist world capacity. In the early 1970s, almost 40 per cent of capacity was either totally owned by the government or held by companies with a majority public ownership.[2] These tendencies are neither limited to developing countries nor to minerals. Production, marketing and exports of internationally traded agricultural commodities like coffee, cocoa, sugar or rubber are also increasingly being controlled by government institutions.

In this set of circumstances the results of economic analyses of commodity markets, based on assumptions of simple profit-maximising behaviour among agents, become doubtful and uncertain. This is because the objective functions and the consequent behavioural patterns of government bodies are likely to differ from those of private firms. Though some hypotheses about likely behaviour of governments involved in primary commodity production and trade have been advanced, they have neither been tested nor fully integrated into analyses of commodity markets. Hence, our understanding of the behaviour of commodity markets in which governments play important roles is incomplete. The experience of uranium in the past decade strongly underlines that one has to be very cautious in applying standard economic analysis for predictive purposes in such markets.

The second lesson is that one should not neglect the producers' potential to impose monopolistic pricing. Such neglect was clearly prevalent among observers of the uranium market in the period prior to the price explosion. Though it was clear to all that the organisation of the industry and the characteristics of its product corresponded to those generally required for producer collusion, the evidence we have surveyed suggests that until 1974 no one seriously considered the possible implications of a full-fledged uranium cartel. Uranium observers were not alone in their lack of foresight and imagination. There was a corresponding neglect among the observers of the petroleum market prior to the rise in oil prices at the end of 1973.

The lesson to be learnt from the uranium and petroleum experiences is that one must always keep the possibility of monopolistic producer pricing in view when studying commodity markets in which structural conditions make producer collusion feasible. This should be so even in cases where, on account of high long-run price elasticities of demand and supply, profitable collusion is feasible only over a limited period of time. For a variety of reasons, producers may discount the future at such high rates that little importance will be attached to the detrimental long-run consequences of monopolistic pricing when profits can be sharply improved in the short run.

Prospects

From Chart 1.1 we see that nominal uranium prices stagnated from the latter half of 1976, at a level between $40-43 per lb. We conjectured in Chapter 7 that $40 (nominal dollars) was the target price of the uranium cartel. In contrast to the relatively stable nominal price quotation, the real price of uranium has experienced a sharp decline in the most recent years. At the end of 1979, the real price was more than 25 per cent below the level of mid-1977.

In several preceding chapters we have hinted at the reasons for the fall in real uranium prices since 1977. The combination of (a) reduced nuclear power expansion, (b) revocation of plans to increase the enrichment tails assay and (c) renegotiation and cancellation of excessive enrichment contracts has led to a sharp curtailment of current and expected future uranium demand. Output has responded briskly to the high price levels since 1975. Additional production on a large scale is expected in the early 1980s in Australia and several African countries. The imposition of supply quotas in 1972/3, when prices were painfully low, was regarded by producers as a survival measure, and was therefore easy to accept. The discipline required for a reintroduction of similar

measures in the 1980s may be far more difficult to muster. The Australian and Canadian governments could of course try to prevent a further
erosion of real uranium prices by slowing down the development of
new mines, or by imposing minimum prices which rise over time to
compensate for inflation. But with the envisaged expansion of production elsewhere, such measures might merely lead to reduced market
shares for the two countries. Under the emergent circumstances, therefore, the cartel's ability to continue its tight control of the market
through the 1980s is in serious doubt. A factor pointing towards
strengthened uranium prices is of course the substantial rise in petroleum prices in 1979, and the ensuing expansion of quasi rents which
uranium producers might try to appropriate.

The imponderables of political action in the uranium market continue to loom large. The warnings formulated in the preceding section
should make one wary about the validity of predictive economic analysis. Keeping this in mind, we contend that under the circumstances
evolving since 1977, the likelihood is that uranium prices in real terms
will be considerably lower in the 1980s than they were in the latter
half of the 1970s.

Our view of the price prospects for the 1980s is based on the premiss
that the uranium market was governed between 1974 and 1979 by an
exceptional set of circumstances which is no longer in effect. In specu-
lating about the levels that the price might assume during the present
decade, it may be instructive to refer back to the price graph of the
introductory chapter, reproduced below as Chart 9.1. We feel that what
was said in Chapter 4 about incentive prices required to attract new
investment capital to the industry continues to have validity. If anything, the incentive price level required to assure today's slower nuclear
expansion needs should be below the $19 level (1975/6 dollars) estab-
lished in 1972. Be that as it may, we have inserted into the price graph
a flat line representing the 1975/6 $19 incentive price, and a step curve
representing the same incentive price in current dollars. While the development of the nominal incentive price during the 1980s will depend on
inflation, the real incentive price will remain at the $19 (1975/6 dollars)
level, unless there are changes in the real costs of marginal uranium
production. It is readily seen that actual quotations at the end of 1979
are far above the incentive price. Barring unforeseen political action
affecting uranium, or serious disturbance in the petroleum and coal
markets, we feel that there is a likelihood that the average uranium
price through the 1980s will not exceed the incentive price level. To be
fulfilled, this prediction requires that the real uranium price fall to $19

Chart 9.1: Uranium Prices, Spot Deliveries, as Reported by NUEXCO
Quarterly Averages

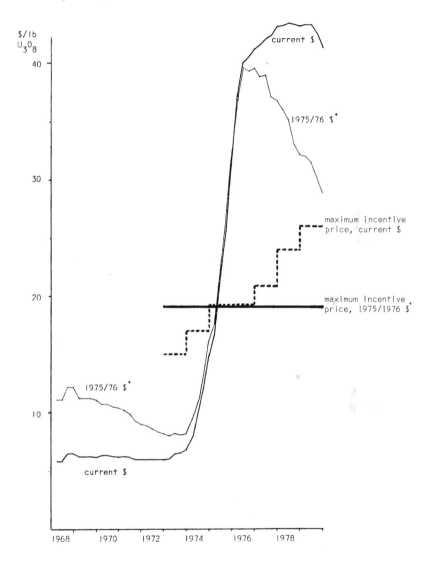

Note:[a] Current dollars deflated by the general dollar GDP deflator for the entire
OECD area. The quarterly GDP deflator data have been obtained through linear
approximations of annual figures (see Table 2.3).

(1975/6 dollars) or below. With continued fast inflation, the prediction could well be fulfilled with nominal uranium prices remaining in the neighbourhood of $40 during the 1980s.

We do not feel competent to make predictions on future political events with repercussions on uranium. Neither can we say what will happen to petroleum. Continued perturbations in the petroleum market, with rising oil prices and uncertain supplies, could certainly have very strong implications for the price of uranium.

On the other hand, we consider it most unlikely that uranium prices would rise during the 1980s on account of resource constraints which necessitated the exploitation of higher-cost uranium deposits. Current low-cost reserves as defined in Table 4.3 are more than adequate for any conceivable crash development programme for nuclear power dring the 1980s. To substantiate this claim we have plotted in Chart 9.2 the Western world uranium requirements as projected by OECD in 1977, and by the Uranium Institute in 1979. These two are then compared with the maximum feasible projection contained in OECD's 1973 survey. This, incidentally, is the highest projection for uranium requirements in the 1980s that we have been able to find. While a nuclear power expansion programme commensurate with this magnitude of requirements might have been feasible if it had been started in 1973, it would hardly be so if initiated in 1980. Uranium requirements in the 1980s must consequently fall substantially below this extreme projection. Yet, it is interesting to note that Western world low-cost reserves as of 1979 constitute 125 per cent of the cumulative ten-year requirements postulated by this projection, and that this is substantially more than the corresponding 1968 figure based on that year's low-cost reserves and realistic requirements forecasts (see Table 4.3). The 1979 low-cost reserves constitute 276 per cent of the requirements for the 1980s as assessed by OECD in 1977, and 357 per cent of the requirements for the same period as assessed by the Uranium Institute in 1979. Even without further additions, the low-cost reserves existing today are more than adequate to cover any feasible uranium requirements well into the 1990s. And there is a virtual certainty that more reserves will be identified as the years go by.

Judgements of the longer-run adequacy of low-cost uranium resources to satisfy future nuclear needs become far more speculative. Much will depend on the growth in requirements which in turn will vary with the growth of nuclear power generation, the development of technology to economise on the use of uranium in light water reactors and with the timing and speed of the introduction of the breeder. The Inter-

Chart 9.2: Western World Uranium Demand Projections

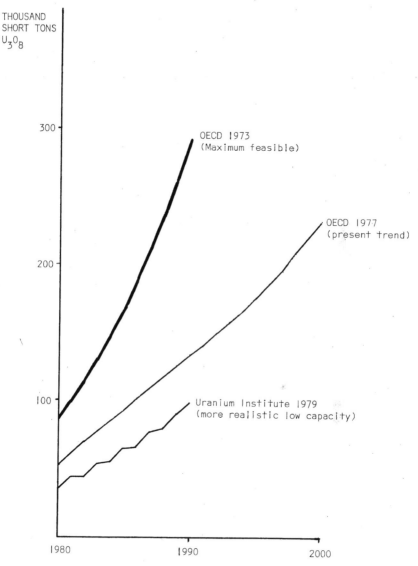

national Nuclear Fuel Cycle Evaluation assesses the cumulative 1980-2000 requirements at between 1.7 and 2.9 million short tons U_3O_8. For the period 1980-2025 the range of 4.6-15.7 million short tons U_3O_8 is given.[3] To compare, we may note that a total of about 0.75 million short tons U_3O_8 was mined in the Western world prior to 1980. It is not purposeful to relate requirements for periods as long as these to identified reserves or resources, because the latter have been established in response to mining companies' planning needs, and are dependent on the spread and intensity of past exploration effort. Especially for a new mineral like uranium, with a very short history of scattered exploration, it may be more relevant to compare the very long-run requirements with assessments of global uranium availability, whether discovered or not, derived from basic geological induction. Table 9.2 lists one such recently published study's estimates of the uranium agglomerations in the earth's crust.

Table 9.2: Estimated Uranium Agglomerations

Grade of U (per cent)	Quantity of U (million short tons)
$>$ 1.0	0.2
0.3 − 1.0	2
0.1 − 0.3	70
0.03 − 0.1	120
0.01 − 0.03	2,000

Source: K.S. Deffeyes and I.D. MacGregor, 'World Uranium Resources', *Scientific American* (January 1980).

One must of course be extremely cautious about deriving conclusions concerning future availability and cost from such figures. Apart from the fact that the quantities given are hypothetical deductions, there is no uniform relationship between grades and costs of recovery. A substantial proportion of the higher-grade agglomerations may prove hopelessly uneconomic because they are too small or situated too deep down in the earth's crust. It should also be noted that the figures comprise the entire world, while the requirements projections are for the Western world only.

Practically all uranium exploitation until the present has been from the three highest-grade categories. The only exception is the large Rossing mine in Namibia, where grades are reported at between 0.03

and 0.05 per cent U.[4] Though no cost estimates have been published for this mine, we may get a feel of the cost levels by noting that the decision to develop it was taken at a time when long-run forecasts for uranium prices (including forecasts formulated by representatives of the company undertaking the investment) were below $18 (1975/6 dollars).

In judging the costs at which the large-scale future requirements of uranium could be satisfied, it is instructive to consider the following.

(1) Geological deductions suggest that high-grade uranium reserves (> 0.1 per cent U) many times greater than those existing at present will be established as the need arises. There is no reason to believe that the cost of exploiting these new reserves would generally be higher, in real terms, than that of current uranium exploitation.
(2) The Rossing uranium deposit is tangential in terms of grade to an extremely large quantity of uranium agglomerations in the earth's crust. The more economical among these should assure global requirements in the very long run, at costs not far above those of the Rossing mine, when present technology is employed.
(3) Technology of extraction will no doubt be improved over the years. Judging from the experience of other minerals with a longer history of exploitation, it is reasonable to assume that at the future time when the need arises to exploit the meagre 0.03-0.01 per cent uranium deposits on a large scale, the costs of doing so will not exceed the marginal cost of present uranium exploitation.

Even though nothing can be stated conclusively about uranium prices in the distant future, the facts and inferences presented here suggest that natural resource availability is unlikely to lead to rising uranium production costs in the very long run.

Notes

1. R.F. Mikesell, *New Patterns of World Mineral Development* (British-North American Committee, London, 1979).
2. R. Prain, *Copper, the Anatomy of an Industry* (Mining Journal Books Ltd, London, 1975).
3. 'Availability of Nuclear Fuel and Heavy Water. Summary', INFCE/WG. 1/16 Rev. 1, 5 November 1979 (stencil).
4. P. Maget, 'Uranium naturel', *Industrie Minérale* (May 1975).

BIBLIOGRAPHY

Adelman, M.A. 'Economics of Exploration for Petroleum and Other Minerals', *Geoexploration*, vol. 8 (1970)

An Analysis of Competitive Structure in the Uranium Supply Industry (Federal Trade Commission, Washington, DC, August 1979)

Annales des Mines (July-August 1976)

Applied Atomics (31 July 1979)

Atkins, W.F., of D.S. Robertson & Associated Ltd 'Production Costs for Three Uranium Properties, Northern Territory of Australia' (December 1971)

Atom (May 1969 and June 1969)

Atomic Energy (April 1976)

Atomwirtschaft (September 1975, December 1977, February 1979)

Australian Atomic Energy Commission *Annual Report 1979*

Bain, J.S. *Industrial Organization* (Wiley, New York, 1968)

The Balance of Supply and Demand 1978-1990 (The Uranium Institute, London, 1979)

Ball, G.R. 'World Uranium Supply and Demand Factors as Applied to an Econometric Model, with Special Reference to Stockpiling', *Neue Technik* (February 1967)

Bell, D.D. 'Nuclear Fuel Resources and Price Trends', *The Canadian Mining and Metallurgical Bulletin* (April 1967)

BP *Statistical Review of the World Oil Industry* (London, 1978)

Brinck, J. 'Uranium Prospects and Problems', *Eurospectra*, no. 1 (1971)

Carmoy, G. de 'Nuclear Energy in France', *Energy Economics* (July 1979)

Chamberlin, E.H. *The Theory of Monopolistic Competition* (Harvard University Press, Cambridge, Mass., 1933)

Clarke, J.S., Searby, P.J., and Mazel, L.C. 'UK Uranium Demand and Procurement Strategy', paper presented at the 1970 Foratom Congress, Stockholm

Combs, G.F., and Patterson, J.A. 'Uranium Market Activity', paper presented at Uranium Industry Seminar Grand Junction, Colorado, October 1978

Commodity Price Trends, 1970 edition, World Bank Report No. EC-166/70

Commodity Trade and Price Trends, 1979 edition, World Bank Report No. EC-166/79

Cooper, D.O., of Getty Oil 'Viable Uranium Exploration and Production Economics', Atomic Industrial Forum Seminar on Uranium, March 1973

Deffeyes, K.S., and MacGregor, I. 'World Uranium Resources', *Scientific American* (January 1980)

District Court of Stockholm, Dept. 7, Legal case T 192-194/75, evidence no. VII. 4.86, Internal letter in the Anaconda Company of 14 June 1974, No. H10, Minutes of cartel meeting in Johannesburg, June 1972, and H17, Minutes of cartel meeting in London, October 1973

Eckbo, P. *The Future of World Oil* (Ballinger, Cambridge, Mass., 1976)

Eklund, S. 'Kärnenergi', *Ymer* (1968)

Euratom Versorgungsagentur *Annual Reports* (1971-6)

Faulkner, P. (ed.) *The Silent Bomb: a Guide to the Nuclear Energy Controversy* (Vintage Books, New York, 1977)

Ferrara, G.M. (ed.) *Atomic Energy and the Safety Controversy* (Checkmark Books, New York, 1978)

Financial Times, 4 September 1976

Gehlin, R., of AB Atomenergi, Stockholm 'Uranium Demand and Supply', paper presented at the 1970 Foratom Congress, Stockholm

General Electric 'Uranium 1973-1985, a Materials Resources Planning Report' (December 1973)

Gilleland, J.E., of TVA Statement at hearings before the Joint Committee on Atomic Energy, US Congress, 17 and 18 September 1974

Government of Canada, 'Policy Statements and Press Releases on Uranium Marketing' (Ottawa, undated)

Grey, A.J. 'Current Australian Uranium Position', *Uranium Supply and Demand* (The Uranium Institute, London, 1977)

Griffith, J.W. *The Uranium Industry, its History, Technology and Prospects* (Dept. of Energy and Mines, Ottawa, 1967)

Hanrahan, E.J., Williamson, R.H., and Brown, R.W. 'US Uranium Requirements', paper presented at ERDA's Uranium Industry Seminar, 19 October 1976

Herfindahl, O.C. *Copper Costs and Prices 1870-1957* (Johns Hopkins, Baltimore, 1959)

Hewlett, R.G., and Duncan, F. *Atomic Shield 1947/1952* (Penn State University Press, University Park, 1969)

Hogerton, J.F. *et al.*, of the S.M. Stoller Corporation 'The Uranium Supply Outlook' (September 1972)

—— 'The Uranium Supply Outlook' (November 1973)

—— 'Report on Uranium Supply' (5 December 1975)

—— 'US Uranium Requirements' in *Uranium Supply and Demand* (The Uranium Institute, London, 1977)

—— 'Disturbing Conflict of Views on Uranium Supplies', *Nuclear Engineering International* (November 1977)

Houdaille, M. 'Le Marché de l'uranium', *Bull. Inform. Ass. Tech. Energ. Nucl.* no. 84 (July-August 1970)

L'industrie minière de l'uranium (Republique Française, Commissariat à l'energie atomique, 1970)

INFCE/WG. 1/16 Rev. 1, 'Availability of Nuclear Fuel and Heavy Water', Summary, 5 November 1979

Jahrbuch der Atomwirtschaft (1977 and 1979)

Jelinek-Flink, P. 'The Impact of Enrichment Policies on the Uranium Market', *Uranium Demand and Supply* (The Uranium Institute, London, 1978)

Johnson, D.M., of United Nuclear Corporation 'Observations on Uranium Prices' (24 November 1975)

Kienlin, A. von 'Commercial Effects of Current Non-proliferation Policies', *Uranium Supply and Demand* (The Uranium Institute, London, 1978)

Klemenic, J. 'Examples of Overall Economics in a Future Cycle of Uranium Concentrate Production for Assumed Open Pit and Underground Mining Operations', AEC Nuclear Industry Seminar, Grand Junction, Colorado, 1972

——, US ERDA 'Uranium Supply and Associated Economics: Fifteen Year Outlook' (October 1975)

Kostuik, J., President Denison Mines Ltd 'Key Issues Affecting the Future Development of the Uranium Industry', *Uranium Supply and Demand* (The Uranium Institute, London, 1976)

Langlois, J.P. 'The Uranium Market and its Characteristics' in *Uranium Supply and Demand* (The Uranium Institute, London, 1978)

Lieberman, M.A. 'US Uranium Resources, an Analysis of Historical Data', *Science* (30 April 1976)

Lloyd, B. 'An Australian View of the Uranium Market', *Uranium Supply and Demand* (The Uranium Institute, London, 1978)

Lorie, J.H., and Gody C.S. 'Economic Analysis of Uranium Prices', a report prepared for Westinghouse Electric Corporation (9 July 1975).

Maget, P. 'Uranium Naturel', *Industrie Minérale* (May 1975)

Mårtensson, M. 'Uran för kraftalstring', *Teknisk Tidskrift* (1967), H 40

——, AB Atomenergi (Sweden) 'Ekonomisk analys av uranmarknadens historiska utveckling' (July 1977)

Merrill Lynch *et al.* Inc. 'The Canadian Uranium Industry, an Institutional Study' (July 1972)

Messer, K.P. 'Uranium Demand as Judged by Electric Utilities' in *Uranium Demand and Supply* (The Uranium Institute, London, 1977)

Metallgesellschaft *Metal Statistics* (1960 and 1977)

Metals Analysis and Outlook no. 2 (August 1976) and no. 4 (September 1977), Charter Consolidated, London

Mikesell, R.F. *New Patterns of World Mineral Development* (British North American Committee, London, 1979)

Mineral Facts and Problems, 1975 edition, US Bureau of Mines, Bulletin 667 (Washington, DC, 1976)

Mining Annual Review (1975-9)

Monthly Energy Review (US DOE) (July 1976 and July 1979)

Mullenbach, P. *Civil Nuclear Power* (The Twentieth Century Fund, New York, 1963)

National Economic Research Associates Inc. 'Competition in Uranium and Coal Markets with Special Reference to Oil and Gas Companies' (New York, June 1979)

Neff, T.L., and Jacoby, H.D. *Nuclear Fuel Assurance, Origins, Trends, and Policy Issues*, MIT Energy Laboratory Report No. MIT-El 79-003 (MIT Press, Cambridge, Mass., February 1979)

Nuclear Engineering (June 1968)

Nuclear Engineering International (November 1977, December 1979 and January 1980)

Nuclear Fuel (25 June, 20 August, 29 October 1979; 21 January 1980)

The Nuclear Industry, USAEC, Wash. 1174 (1971-4)

Nucleonics Week (17 July 1975; 9 September, 21 October, 23 December 1976; 23 June 1977)

NUEXCO Market Report (1972-9)

NUEXCO *Significant Events in the Uranium Market 1969-1976* (October 1976)

Nukem Market Review (1972-9)

Nye, J.S. 'Balancing Non-proliferation and Energy Security', *Uranium Demand and Supply* (The Uranium Institute, London, 1978)

Oboussier, F. 'Supplying the Six', *Nuclear Active* (July 1975)

—— 'Die Versorgung der Bundesrepublik Deutschland mit angereichertem Uran', *Gluckauf*, no. 9 (1976)

—— 'Statement on the World Needs for Uranium, its Demand and Production Possibilities' (Bonn, September 1977)

OECD *Uranium Resources* (OECD, Paris, December 1967)

—— *Uranium Production and Short-term Demand* (OECD, Paris, January 1969)

—— *Uranium Resources, Production and Demand* (OECD, Paris, September 1970)

—— *Uranium Resources, Production and Demand* (OECD, Paris, August 1973)

—— *Uranium Resources, Production and Demand* (OECD, Paris, December 1975)

—— *Uranium Resources, Production and Demand* (OECD, Paris, December 1977)

—— *Energy Prospects to 1985* (OECD, Paris, 1974), vol. 1

—— *Energy Statistics 1975/1977* (OECD, Paris, 1979)

Patterson, J.A. 'Uranium Market Activities' (US ERDA, Washington, DC, October 1976)

—— 'US Uranium Supply and Demand Overview', paper presented at a meeting of the American Nuclear Society, 24 January 1977

Petrick, A.L. 'An Economic Analysis of the Relationship Between the Cost and Price of Uranium' (Petrick Associates, 1979)

—— 'An Economic Analysis of Uranium Prices' (draft, 1979)

Pickard Love & Garrick Inc. 'Natural Uranium, Demand, Supply and Price' (February 1977)

Pickering, J.F. *Industrial Structure and Market Conduct* (Martin Robertson, London, 1974)

Pittsburgh Post Gazette, 6 January 1977

Pottier, M.M. 'Mechanisms of the Uranium Market' in *Uranium Supply and Demand* (The Uranium Institute, London, 1978)

Prain, R., Chairman Roan Selection Trust 'The Future Availability of Copper Supplies', speech at the Institute of Metals, Amsterdam, 1970

—— *Copper, the Anatomy of an Industry* (Mining Journal Books Ltd, London, 1975)

'Proposed Post Trial Findings of Westinghouse in US District Court for Eastern District of Virginia', MDL DOC No. 235 (1978)

Radetzki, M. 'Metal Mineral Resource Exhaustion and the Threat to Material Progress, the Case of Copper', *World Development* (February 1975)

—— 'The Potential for Monopolistic Commodity Pricing by Developing Countries' in G.K. Helleiner (ed.), *A World Divided* (Cambridge University Press, Cambridge, 1975)

—— 'Market Structure and Bargaining Power; a Study of Three International Mineral Markets', *Resources Policy* (June 1978)

—— 'Will the Long-run Global Supply of Industrial Minerals be Adequate? A Case Study of Iron, Aluminium and Copper' in Christopher Bliss and Mogens Boserup (eds.), *Economic Growth and Resources,* Vol. 3, *Natural Resources* (Macmillan for the International Economic Association, London, 1980)

Radetzki, M., and Zorn, S. *Mineral Processing in Developing Countries,* UNIDO/IOD, 328 (UNIDO, Vienna, 19 December 1979)

Robertson, D., & Associates Ltd 'Uranium Supply and Demand', paper prepared for the Advisory Committee on Energy, Province of Ontario, Canada, 1972

—— 'Evaluation of Allied Nuclear Corporation' (July 1974)

—— 'Uranium Price Movement and the Reasons Therefor' (31 August 1977)

Rolph, E.S. *Nuclear Power and the Public Safety* (D.C. Heath, Lexington, 1979)

Rotsay, W.B. 'Conditions Applying to Australian Uranium Exports', *Atomic Energy* (April 1976)

RTZ *Annual Reports* (1975-9)

Runnals, O.J.C., Dept. of Energy, Mines and Resources, Canada, 'Uranium', in *Geos* (Fall 1972)

Scherer, F.M. *Industrial Market Structure and Economic Performance* (Rand McNally, Chicago, 1970)

Shaw, K.R. 'Capital Cost Escalation and the Choice of Power Stations', *Energy Policy* (December 1979)

SIPRI Yearbook 1979 (Taylor & Francis Ltd, London, 1979)

Statistical Data of the Uranium Industry (USERDA, Washington, DC, January 1975, and USDOE, Washington, DC, January 1979)

Styrmedel för en framtida energihushållning (Swedish Ministry of Industry, Os 1 1977:16)

Svenke, E. 'Uran – globala aspekter på råvaruförsörjningen för kärnkraft', lecture delivered at the Swedish Geological Society symposium, 28 March 1974

Swedish State Power Board Memo. ER-J11/KW-6320, dated 7 October 1971

—— contract with Uranex of 13 June 1973

—— memo. of 4 May 1977, signed by I. Wivstad

Technique de l'Energie (June 1979)

Uranium Price Formation, EPRI (October 1977), prepared by Charles Rivers, Cambridge, Mass.

Uranium Supply and Demand (The Uranium Institute, Mining Journal Books Ltd, London, 1976, 1977 and 1978 issues)

US AEC, GJO-100 (January 1976)

US Mineral Resources Geological Survey, Professional Paper 820 (Washington, DC, 1973)

Van Duyne, C. 'Commodity Cartels and the Theory of Derived Demand', *Kyklos*, vol. 28 (1975)

Westerhaugh, O., and Blomsnes, B. 'Trends in Nuclear Power Costs in Sweden', *Nuclear Engineering International* (December 1979)

White, G., of Energy Services Company, a division of NUEXCO 'Ten Year Forecast of Price Trends in the Domestic Uranium Industry' (July 1972)

Williams, R.M. 'Uranium Supply to 2000, Canada and the World', paper presented at a meeting of the Geological Association of Canada, Edmonton, Alberta, May 1976

—— 'Uranium', *Canadian Mining Journal* (February 1978)

Worrol, R.E. 'Uranium Mining and Marketing in South Africa' in *Natural Uranium Supply* (International Symposium of Deutsches Atomforum, Mainz, November 1974)

—— 'The Pattern of Uranium Production in South Africa', *Uranium Supply and Demand* (The Uranium Institute, London, 1976)

INDEX

Printed and bound by CPI Group (UK) Ltd, Croydon, CR0 4YY

22/10/2024

01777628-0017